CONTROL OF ARTHROPODS
of Medical and Veterinary Importance

CONTROL OF ARTHROPODS

of Medical and Veterinary Importance

Edited by

R. Pal

World Health Organization
Geneva, Switzerland

and

R. H. Wharton

Division of Entomology, CSIRO
Long Pocket Laboratories
Brisbane, Queensland
Australia

PLENUM PRESS • NEW YORK AND LONDON

Library of Congress Cataloging in Publication Data

Symposium on Control of Arthropods of Medical and Veterinary Importance, Canberra, Australia, 1972.
Control of arthropods, of medical and veterinary importance.

Held at the 14th International Congress of Entomology, Canberra, Australia, August 22-30, 1972.
Includes bibliographies.
1. Fly control—Congresses. 2. Mosquito control—Congresses. 3. Tick—Control—Congresses. 1. Pal, R., ed. II. Wharton, R. H., ed. III. International Congress of Entomology. 14th, Canberra, Australia, 1972. IV. Title. [DNLM: 1. Arthropod vectors—Congresses. 2. Communicable disease control—Congresses. AZ460 C764 1972]
RA641.T5S9 1972 614.4'32 74-4172
ISBN 978-1-4684-2093-7

ISBN 978-1-4684-2093-7 ISBN 978-1-4684-2091-3 (eBook)
DOI 10.1007/978-1-4684-2091-3

Proceedings of the symposium "An Assessment of the Current Status and Prospects for the Control of Arthropods of Medical and Veterinary Importance," held at the 14th International Congress of Entomology, Canberra, Australia, August, 1972

Softcover reprint of the hardcover 1st edition 1974

A Division of Plenum Publishing Corporation
227 West 17th Street, New York, N.Y. 10011

United Kingdom edition published by Plenum Press, London
A Division of Plenum Publishing Company, Ltd.
4a Lower John Street, London W1R 3PD, England

CONTENTS

INTRODUCTION

R. Pal

World Health Organization

Geneva, Switzerland

The 14th International Congress of Entomology was held at Canberra from 22 - 30 August 1972. The programme of the Congress was divided into 21 symposia and 15 sections. One of the symposia dealt with an assessment of the current status and prospects for the control of arthropods of medical and veterinary importance. Eleven papers dealing with houseflies, ticks, blackflies, **Aedes aegypti** *and other* **Stegomyia** *species, tsetse flies,* **Culex tritaeniorhynchus** *and* **Culex fatigans** *were presented. These papers are published herein with very slight changes along with the discussions which followed.*

During the past 25 years, spectacular success has been achieved in the control of vector-borne diseases such as malaria, plague and yellow fever. This has been possible due largely to the use of DDT and other residual insecticides at a cost within the reach of developing countries where these problems were most prevalent. Despite the remarkable successes achieved, the recent epidemics of malaria in Ceylon, typhus in Burundi, plague in Viet-Nam, and yellow fever in Ethiopia and Ghana have given rise to a sense of pessimism. This pessimism is due not only to the recrudescence of vector-borne disease in some areas, but also to the fact that there are still no organized programmes for the control of Chagas' disease, onchocerciasis and schistosomiasis. Thus the need for vector control, particularly in developing countries, remains undiminished.*

** Since this Congress, an organized programme for onchocerciasis control has been started in West Africa (see Chapter 6).*

The problem has become much more complicated since the introduction of DDT and other organic insecticides because of the development of resistance of insects to insecticides and the environmental pollution associated with the widespread use of pesticides. At present, vector control is almost entirely achieved by chemicals, and as far as can be judged, insecticides will remain the principal weapon of attack during the immediately foreseeable future. Increasing emphasis is being placed on the development of alternative methods of control, such as biological control, insect growth regulators and pheromones, chemosterilants, and genetic control techniques. The methods are relatively sophisticated, and at present none of them is operational.

The papers presented at the symposium have clearly shown that the future of vector control will depend on the integration of various methods instead of on one method alone. A series of methods complementary to each other may be employed: for example, an initial suppression of the population by means of environmental sanitation and pesticides, followed by biological and genetic control to suppress the residual population. With the use of chemicals alone, there usually remains a residuum of the vector population, with the result that the chemicals have to be used year after year, giving rise to the problems of insecticide resistance and environmental pollution. In fact, there are already instances of vector species including Culex tarsalis, Aedes nigromaculis, Musca domestica *and ticks, which have developed resistance in some areas to all the three chemical groups of insecticides currently available.*

In some instances this has necessitated a return to older methods of control such as the use of Paris green or larvicidal oils. Perhaps this has led Dr McClelland to remark that "recent doubts have been cast on the prospects of eventual communicable disease control, and it may be that man must turn from expensive technology back to earlier and cheaper but nevertheless effective methods."

One other point which might be mentioned is the current controversy over control versus eradication. The eradication programmes involve greater capital expense and superhuman effort which make the consequences of recrudescence all the more disastrous, as have been exemplified by malaria, yellow fever or Aedes aegypti *eradication projects. So long as the density of vector species can be reduced to a critical level at which it can no longer transmit the disease, the major objective can be achieved. Whether the objective is to eradicate or control a disease vector, sustained effort is required.*

It has also become clear that constant surveillance of vectors and vector-borne diseases should be maintained on a global basis and any outbreak of disease should be immediately dealt with by emergency epidemic control measures such as ultra low volume application of an appropriate pesticide. The World Health Organization has already taken the initiative in this direction by stock-piling insecticides and equipment at strategic places throughout the world and making a staff of consultant experts available to plan and direct emergency control programmes. The reader is referred to a special publication on vector control issued by the World Health Organization (WHO Chronicle Vol. 25, No. 5, 1971). The chapter on the future of vector control ends with the following words:

"All these control methods - genetic, biological and chemical - will be integrated in order to achieve maximum control at minimum cost. In this way it is hoped that an increasing human population will live safely in the midst of a vector population reduced to innocuous levels."

We may, therefore, begin the symposium with this optimistic note.

HOUSEFLIES (*MUSCA DOMESTICA*)

J. Keiding

Danish Pest Infestation Laboratory

Skovbrynet 14, DK-2800 Lyngby, Denmark

INTRODUCTION

The assessment of the current status and prospects for the control of houseflies is partly based on the literature, but primarily on 30 years of personal experience in housefly control in temperate Denmark and on visits to many parts of the world, including tropical and subtropical areas, where I observed and discussed control problems.

The housefly, *Musca domestica*, occurring in several subspecies all over the world is dependent on organic waste from human activities as a breeding source, primarily garbage from households and industry and the excrement of domestic animals. Where such breeding sources are not adequately reduced and the climate is suitable, high fly populations will occur.

CHEMICAL CONTROL

(Concerning insecticides and control methods see Keiding 1964; WHO 1970 and CDC 1972)

With the advent of synthetic insecticides, effective chemical fly control became possible, primarily by residual treatment of surfaces where flies aggregate. DDT was the "miracle" insecticide for fly control immediately after World War II. However, within two to three years houseflies developed a high level of resistance to this and other chlorinated hydrocarbons. In the 1950's, effective organophosphorous (OP) insecticides became available, and housefly control with residual sprays, treated cords and strips, or baits was successful in many areas. Control failure due to the development of resistance to the newer compounds was first noticed in Denmark and Florida in 1955, (Keiding 1956; Lindquist 1956) and later in many other areas (for review up to 1969 see Brown and Pal,1971, p. 255-302).

However, by replacing one type of OP-compound with another, it was possible until the latter half of the 1960's to achieve reasonable fly control by the use of residual OP-compounds, even in the most difficult areas in Scandinavia and the United States of America. In the last few years, the point has been reached in some places where no insecticides are effective for residual fly control. Examples of successive failures of insecticides due to resistance are: (1) diazinon,(2) malathion, (3) ronnel, (4) fenitrothion, (5) fenthion, (6) trichlorfon, (7) dimethoate and (8) dichlorovos, involving serious cross- or co-resistance to several other OP-compounds and to carbamates. On Danish dairy and pig farms, significant resistance to some compounds was developed within two to three seasons to (1), (2), (3) and (4) (Keiding 1965b, 1967), whereas a longer period was required for the others, especially (7), where the residual effect of 1 g/m^2 did not generally fail until the sixth or seventh year of widespread use (Keiding, 1972, see Table 3). How serious the situation may become is apparent from Table 1, which presents data on resistance to various OP-compounds on Danish farms, in 1971.

Areas with similar, but not quite as extreme, resistance problems are dairy farms and chicken factories in the United States (e.g. California: Georghiou 1966; Georghiou and Bowen 1966; Georghiou and Hawley 1971, see fig. 1; Georghiou et al.,1972. Florida: Brady et al.,1966 ; Bailey et al.,1967, 1970a, b; CDC 1972. Georgia: Mathis et al.,1968 and 1972, see Table 2; New Jersey: Hansens et al.,1967, 1968 and 1970, Table 2). It is interesting to note the low to moderate resistance to Gardona and the carbamate Mobam and their fair effectiveness in these problem areas. In Denmark, Mobam in 1967, and Gardona in 1969, failed completely as a result of high resistance already present (Keiding 1968, 1970). The factors favouring the development of resistance seem to be prolific fly breeding and continuous insecticidal pressure in the fly season to semi-isolated populations.

In other areas with a long history of chemical fly control, the problems seem to be less serious. For instance, recent tests from Japanese farms showed serious resistance only to malathion and diazinon, and moderate or low resistance to fenitrothion and bromophos (Hayashi et al., 1971, 1972). In the Latina area in Italy, Sacca et al. (1970) found low tolerance to 15 OP-compounds and only moderate resistance to trichlorfon and Dursban after 15 years of fly control first with diazinon and, later, with fenthion and ronnel. In Ohio dairy barns, the effect of dimethoate did not

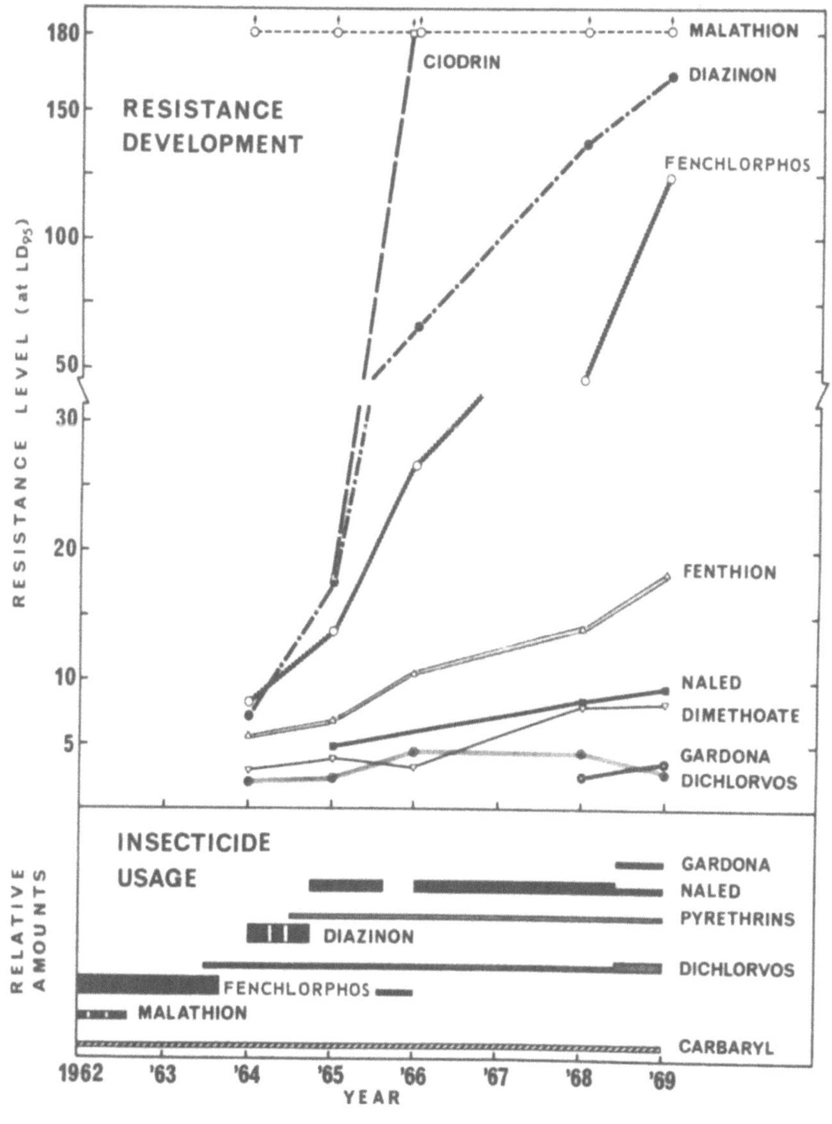

Figure 1. Changes in resistance ratios in houseflies 1964-69, and the usage of various insecticides 1962-69 at a poultry ranch in Southern California (From Georghiou et al. 1971) (Note that the scale of resistance level on the ordinate is equidistant except between 30 and 50; usually such scales are logarithmic.)

decrease during five years of use, and good control was obtained in 1971 with several other OP-compounds including fenthion, jodfenphos and Gardona, but not with diazinon and bromophos (Treece et al.,1972). In the Soviet Union chlorofos, a technical trichlorfon, has been used widely for 10 to 15 years without serious resistance developing (Gvozdeva et al., 1972, Lineva 1970 and 1972). This may be due to less selection pressure and/or to the fact that the treatments act partly as stomach poisons. Trichlorfon is a peculiar material: in Denmark it was used as a residual sugar bait for many years without resistance problems, but these developed rapidly and generally as a result of strong selection pressure with other OP-compounds, particularly fenthion and dimethoate, used as residual sprays (Keiding 1965b,1968, 1969). In Florida, rather high resistance developed to the stomach-effect of trichlorfon, but it was still quite effective (Bailey, et al.,1970a). This may be due to the effect on reproduction as reported from the Soviet Union (Lineva 1966, 1972, 1970 and Balan 1970).

Finally, there are still many areas in the world with less intense or organized fly control, where high resistance has been found to chlorinated hydrocarbons only and perhaps to one or two OP's, e.g., malathion and diazinon. Examples are Egypt (Toppozada et al.,1967; El-Deeb et al.,1969; Enan 1970; Abdel-Gawaad et al., 1972), tropical Africa (Keiding 1972),Thailand (Kerdpibule et al., 1971), Australia (Millar et al.,1969), Canada (Bath et al.,1970), and Europe (Sztankay-Gulyas et al.,1970).

Prospects of Housefly Control with Insecticides

There are two situations to be considered:

(1) Resistance has made all suitable residual insecticides ineffective (as in Denmark and parts of the United States).

(2) Some effective residual insecticides are still available.

(1) The prospects of new effective OP insecticides appear to be very gloomy. The multi-resistance in Denmark combines several resistance mechanisms, e.g. various esterases, microsomal oxidations, and reduced penetration (Sawicki et al.,1968; Sawicki 1970a and b, 1972; Lewis and Sawicki 1971). Against any new type of OP-compound tried since dimethoate, Danish flies show some and often considerable resistance, which quickly increases when the fly populations are exposed to insecticide pressure. Carbamates offer even less hope due to the high resistance present before their use, and the chlorinated hydrocarbons were of course abandoned long ago. Concerning pyrethrum and the pyrethroids, so far

there is none with a reasonable residual effect. If there were such a formulation, resistance would probably soon appear. It is true that there has not been much resistance to synergized pyrethrums and pyrethroids as yet, but this is probably because the selection pressure has not been strong for a sufficiently long period. It is known that houseflies have the ability to develop such resistance (Davies et al.,1958; Keiding 1972). Really new types of insecticides are needed but none are yet known with a residual contact effect at a reasonable price.

In recent years, it has been found in laboratory tests that resistance to several OP-compounds, especially dimethoate, fenthion and trichlorfon, is overcome by high dosages of certain synergists that inhibit microsomal detoxifying enzymes (unpublished results referred to in Yasutomi and Keiding 1969; Keiding 1970 and 1972). The effect of many carbamates are also greatly increased by these synergists. It is possible that residual insecticide-synergist mixtures, effective against resistant flies will become available in the future, but there are several problems involved (Oppenoorth 1971; Wilkinson 1968, 1971): the cost of the synergist, which must be used at relatively high ratios, the mammalian toxicity of the mixture, and the presence of other resistance-mechanisms.

Space sprays or aerosols with synergized pyrethrum or pyrethroids can still kill flies in most places. They are useful for temporary relief indoors, and frequent use in piggeries and cowbarns may give satisfactory control on the farm if fly breeding is well reduced (Funder 1972). Limited trials with automatic aerosol dispensers have been promising, but the method is expensive, needs good maintenance and may lead to the development of pyrethrum-resistance.

Short-term fly control out-of-doors can be obtained by mist spraying or ultra low volume sprays, with various OP-compounds and pyrethroids (Kilpatrick et al.,1963; Davies 1966, 1971; Lofgren et al.,1970; Axtell 1971; CDC 1972). This may be useful at refuse dumps and for emergencies.

In toxic baits there are greater possibilities of compounds to counteract resistance. Their use, too, requires reasonable sanitation, and reduction of breeding and food for flies; and their effect is limited by the lack of strong chemical attractants. The development of an attractant would be a breakthrough in housefly control, but there has been little real progress in recent years. The difficulty is (1) that the attractant must compete with the natural attractants present in the area to which local fly populations have become adapted and (2) that *Musca domestica* is poorly equipped with smell receptors (Wiesman 1960, 1962) and finds its way by a high degree of exploratory activity combined with orientation by vision, taste and smell at short distances. However, feeding stimulants and attractants occurring, for example, in hydrolysed or decomposed yeast and animal protein (Mer et al.,1960; Robbins et al.,1965; Neselovskaya et al.,1968; Privora et al., 1969; Volcsik 1971) ammonium carbonate, syrups (Sehgal et al., 1970) or malt extract may be useful to increase the effect of toxic baits. A sex pheromone, muscalure, has recently been isolated from female houseflies and synthesized in the United States, but its attractant power for male flies is not impressive (Carlson et al., 1971). Field tests in Florida have shown that muscalure may greatly enhance the effectiveness of baits or traps for houseflies (Carlson et al., 1973).

Selective Fly Control

Some research workers have tried to make chemical control of adult flies more selective by detailed studies of the distribution of flies within an infested premise and the factors determining movements, attraction, preference and rest, by day and by night (Maier et al.,1952; Kilpatrick et al.,1952; Anderson et al.,1964; Keiding 1965a; Groth et al.,1970). Our investigations in piggeries demonstrated that a high proportion of the flies would rest on the upper third of the partitions between the pens, and if this area, only 1/10 of the total sprayable surface is treated with an effective insecticide, good fly control can be obtained (ben Hannine and Keiding 1967-1968; Keiding and ben Hannine 1967, and 1972). We call this strategic or selective fly control. The reduction in the amount of chemical applied means that more expensive or more toxic formulations may be used, such as strips of contact- or stomach-poisons applied with a brush. In many areas the houseflies strongly prefer to rest at night on edges, wires, cords, tapes, etc. on or hanging down from the ceiling, and this preference lends itself well to strategic fly control, either by selective spray treatment or the use of impregnated cords, tapes, paper, etc. (Wichmand 1953; Kilpatrick et al.,1956; Döhring,1966; Varma et al.,1966-1967).

In areas with relatively warm nights, many houseflies rest at night in vegetation. A study of such outdoor resting sites may be used for strategic fly control, or flies may be lured to rest on artificial resting surfaces based on studies of housefly resting habits. In recent years at our laboratory (Arevad 1972, 1968-1969 and unpublished), we have especially analyzed the movements and preferences of newly emerged flies until the time they can use their wings. The results may be useful for strategic fly control.

When strategic fly control is carried out with suspended impregnated materials or bait treatment, resistance seems to develop more slowly than when residual contact sprays are used, but the effect may not be immediate, or to put it another way: the need for reducing fly breeding will be greater.

Larvicides

(for recent review, see Miller 1970)

So far only insecticides against adults have been discussed. In theory larvicides appear important, but in practice they have several limitations. They must be applied regularly, often at intervals of a few days, as larval medium is continuously accumulated; they usually destroy natural enemies; and resistance to the OP-compounds that are effective may develop even more rapidly than with adulticides. However, with larvicides there are more possibilities of using chemicals with new types of action, e.g. juvenile hormone action (Wright et al.,1972) or interference with moulting (Mulder 1972), than with adulticides, but the economical and practical aspects have not been favourable as yet. Many investigations, mainly in the United States (see Miller 1970; Miller et al., 1972) have shown that it is possible to include larvicides in the feed of cattle and poultry, so that the manure becomes toxic to susceptible fly larvae, with no detrimental effect on the animals and without a residue left in meat, eggs or milk. Yet it is doubtful whether this systemic application will become important in fly control with the growing objection to any chemical contamination of food, apart from the problem of resistance to the OP-compounds used.

In the areas in which the insecticides useful for residual treatments have not been exhausted, the main reason is probably that the insecticidal pressure on the fly population has been less strong and consistent than, for instance, on the Danish or American farms. In order to develop a strategy for using insecticides in such areas so as to avoid or greatly delay general multi-resistance, the basic principles involved are: to reduce insecticidal pressure; to emphasize source reduction and to apply insecticides only as a supplement and in a strategic way both as to location and timing; to avoid widespread residual treatments, perhaps especially with dimethoate, and to rely more on baits with stomach poisons (including trichlorfon) as in the Soviet Union and non-residual sprays (pyrethrum, etc.) where feasible. Alternating between two or more insecticides with different chemical structures is another possibility, although unfortunately most attempts to date have been disappointing. Resistance to the alternating insecticides usually demonstrates some positive correlation even in cases where the main specific resistance mechanisms (e.g. detoxifying enzymes) are different, and high resistance, especially to OP-compounds, is often dependent on interaction between unspecific and specific mechanisms. Nevertheless, alternation is worth considering, especially at an early stage before resistance to the compounds involved has developed to a high degree. When insecticidal pressure with a particular insecticide is relaxed, some reversion to susceptibility may occur, but the redevelopment of resistance becomes more rapid when its use is resumed, and the longer pressure is exerted with it, the more stable the resistance becomes (Keiding 1967).

BIOLOGICAL CONTROL AND REGULATION OF FLY POPULATIONS

In habitats that are not too temporary or unstable, a rich variety of arthropod predators which feed on housefly eggs, larvae or pupae may be found (West,1951; Legner & Olton 1968b, 1970; Peck et al.,1969; Peck,1969), and several species of parasitoid hymenoptera are specialists on fly pupae (Legner & McCoy 1966; Legner et al.,1967; Legner & Olton 1968a; Legner & Greathead,1969; Bennett & Legner,1967). Under relatively stable conditions and in a suitable climate, these fly predators may limit the fly population to an acceptable level. This is particularly the case in tropical areas, where ants seem to play an important role (e.g. noticed in Malaysia (Reid,1953), the West Indies (Pimentel 1954), and the Philippines (Pimentel et al.,1969)). Other important predators are various beetles, fly larvae and macrochelid mites (Axtell 1963a, b, c and 1969; Rodriguez 1972; Eröss et al., 1971) particularly *Macrocheles muscaedomesticae* (Singh et al.,1969). The latter are especially important in cool climates. In habitats where regulating organisms may exist, it is important to preserve them, make conditions favourable for them and certainly not kill them with larvicides or other insecticidal treatment. In other words, predators should be an important part of integrated control (Predator mites: Axtell 1968, 1969, 1970a, b; Singh et al.,1969; Wicht & Rodriguez 1970. Other predators or parasites: Legner & Brydon, 1966; Legner & Olton 1968b; Legner,1971; Peck et al.,1970; Mourier 1972).

Active biological control by release (inundation) of parasitoids or predators in the USA (Legner & Brydon 1966) and in Denmark (Mourier 1972) have not been very effective, and it is doubtful if it will become important except in certain isolated areas. More hope may perhaps be attached to developing effective fly diseases, fungal such as *Enthomopthera*, bacterial or viral, but little has been done so far. This will require development of more virulent strains than normally found, methods of mass-production, storage and application, the problems common to all microbial insect control.

In recent years there has been a great interest in using the sterility methods for fly control. Many chemosterilants are effective at low dosages, but safe economic compounds are not yet available. Small scale field tests particularly in the United States have been promising (The literature on chemosterilants is very extensive; for a recent comprehensive review see Campion 1972. Concerning field tests with houseflies see Weidhaas 1968, La Brecque et al.,1966; Meifert et al., 1967, 1969, 1971; Lancaster et al.,1969; Pausch 1972; Loaeza et al., 1965; Sacca et al., 1966,

Duport et al.,1968, 1972; Matsuzawa et al.,1968; Zakharova 1971). If the risk to man and animals can be sufficiently eliminated, chemosterilants may become an important weapon in the battle against houseflies, especially if better attractants were to be developed.

Theoretically release of mass-reared sterilized flies may be used effectively in integrated control schemes, as shown by the calculations of LaBrecque and Weidhaas (1970) and in studies on the potential rate of increase in natural populations of houseflies (Weidhaas & LaBrecque 1970; LaBrecque et al.,1972). However, a high level of organization and expertise is required for such a scheme; there is usually an objection to the temporary increase of the housefly numbers, and the natural fly population must be low at the time of release, either due to seasonal reduction (in winter and spring) or due to control by chemicals or other means. Field experience on the release of sterile houseflies is very limited (Magaudda et al.,1969; Mourier 1973). Except for well-isolated, small populations the eradication of houseflies by the sterility methods seems unrealistic.

Similar problems occur in genetic fly control, unless a self-propagating effect is available. However, there are highly interesting theoretical possibilities for fly control by release of suitable karyotypes (chromosome translocations) and genotypes, where both sterility, conditional lethal effect and genetic displacement (e.g. to restore insecticide susceptibility) might work together (Whitten 1970, 1971; Bushland 1971; Foster et al.,1972; Smith and von Borstel 1972). The potential of genetic manipulation of insect populations was discussed by the experts in other symposia and sections at the Congress. The methods require extensive fundamental studies on the genetics of the species, including translocation, mutation, etc. such as the current work by Wagoner et al., US Department of Agriculture on *Musca domestica*.

TRAPPING

Lack of success with insecticides for fly control has also revived the interest in using traps. Traps and "electrocutors" with black-light or other light as attractants are frequently tried (Morgan et al.,1970; Thimijan et al.,1970; Tarry et al.,1971), but their ability to lure houseflies and reduce populations is rather limited (Thimijan 1972; Funder 1973) and very dependent on location, temperature and the physiological condition of the flies (sex, age, maturity, nutrition, etc.) (Pickens et al.,1969; Hecht 1970). The effect of baited traps is limited by the lack of a really strong chemical attractant, as mentioned previously. Therefore, unless better attractants are found, traps will not be important in fly control under most conditions.

SANITATION AND SOURCE REDUCTION

The prospects of fly control by chemical and biological methods have now been considered and not found very promising, and so there must be a return to the original concept that control of houseflies is primarily a question of eliminating the breeding possibilities.

Technically this is possible everywhere, and the simple principles were developed and described 40 to 70 years ago (see discussions and reviews in Patton 1931; West 1951; Johnson 1959). Such measures include: for animal excrements collection of manure and composting or covering of manure heaps to kill larvae and pupae by fermentation heat (e.g. Austen 1926; Roubaud 1936; Thomsen 1934; Gotaas 1956; Ascher 1961; Sento 1961, 1965; Omori et al.,1962; Eastwood et al.,1967; Dobson et al.,1970), spreading and drying in dry climates, installation of fly proof latrines (Wagner et al., 1958; Ross Institute 1968); and for garbage: systematic storage, collection and disposal of organic refuse from households, shops, markets, industry, farming, etc. (see handbooks on waste collection and disposal).

In cities and towns with a reasonable organization (urban areas) and economy,there should be no excuse for flies having opportunities to breed. Apart from economy,it is a question of organization, collaboration and education at all levels from the housewife to the city administrator. In rural areas it may be more difficult to eliminate fly breeding sources, due to lack of organization, little understanding of the problems and no interest in solving them, because the sanitation and living standard is low in other respects. The most difficult rural housefly problems occur where animal dung or other organic matter is mixed with moist soil and form extensive breeding grounds (Examples: villages in Egypt (Holway et al.,1951), vegetable fields in South-east Asia (Reid 1953; Wharton et al.,1962, personal observation), feedlots in the United States (Walsh,1964)). In developing countries, the sanitary solutions to the fly problems may be opposed because of old habits and traditions, whereas in developed countries opposition is often due to the lack of labour and interest (e.g. in connection with production of chickens, pigs, cattle and in the food industry).

Having realized that source reduction must be the basis of fly control and that the methods are known and feasible, perhaps the most important task in fly control now is to sell this idea, adapt the sanitary anti-fly methods to the various economic, social and cultural situations, make them not only acceptable, but desirable, both technically and socially.

Table 1

INSECTICIDE-RESISTANCE IN F_1 GENERATION OF HOUSEFLIES COLLECTED ON DANISH FARMS IN 1971[1]

Topical application tests on females

Insecticide tested	Fly control treatments[3]		Resistance ratios (R/S) at LD 95 1971	
	1970	1971	Pre-treatment (no)	Post-treatment (no)
DIMETHOATE	dimethoate dimethoate dimethoate	dimethoate jodfenphos pyr/p.b.Aer.	31-305 (9)	139-350 (7) 48-137 (2) 34- 180 (4)
JODFENPHOS[2] FENTHION FENITROTHION	various OPs dimethoate dimethoate	jodfenphos dim. or jod. dimethoate	10-26 (4)	17-77 (4) 38-52 (2) 37-64 (2)
PYRETHRIN/ P.B. 1:5	dimethoate or di-chlorvos	pyr/p.b.Aer.	2.7-4.4 (14) 2.6 (1)	2.5-7.2 (14) 23 (1)

[1] To the results given in the table should be added: General high resistance to diazinon and very high (R/S at LD 95 100-300) to malathion, tetrachlorvinphos (Gardona), trichlorfon and Mobam (a carbamate).

[2] Cross-resistance to bromophos and ronnel.

[3] Residual sprays, aerosols (Aer.) or residual fumigant (dichlorvos).

no = Number of farms investigated

p.b. = piperonyl butoxide

Table 2

RESISTANCE RATIOS (R/S) TO VARIOUS ORGANOPHOSPHORUS COMPOUNDS IN SELECTED FARMS FROM THREE AREAS IN THE UNITED STATES, WHERE INTENSIVE CHEMICAL FLY CONTROL HAVE BEEN CARRIED OUT FOR MANY YEARS

	A	B		C			
	R/S at LD 50 in houseflies from 10 dairy farms in New Jersey. Topical application in F_1 (Hansens et al. 1970)	R/S at LD 95 in houseflies from a typical dairy farm at Savannah, Georgia. Dipping test of field collected flies (Mathis et al. 1972)		R/S at LD 95 in houseflies from poultry ranches (P) and dairies (D) in Southern California. Topical application in F_1 - F_4 (Georghiou et al. 1972)			
	1968	1969	1970	1970		1971	
				P (5)	D (1)	P (2)	D (4)
Diazinon	8 - > 25			12 - 580	26		
Ronnel	10 - 35	21	22	12 - 737	19		
Bromophos	8 - 18						
Jodfenphos		14					
Fenthion				5.8 - 43	8		
Dimethoate	13 - 48	21	69	4.1 - 15	15	4.8 - 8.4	38 - 105
Naled			15	2.4 - 26	3.3	16 - 26	11 - 14
Dichlorvos				1.0 - 7.9	1.3		
Gardona	8 - 17	13	10	1.6 - 4.2	1.4		

Table 3

LEVELS OF RESISTANCE TO DIMETHOATE IN HOUSEFLY POPULATIONS ON DANISH TRIAL FARMS 1963-71+

(Topical application in F_1)

Year	No. of farms	Per cent of trial farms with pre-treatment R/S (at LD 95)							3 highest pre-treatment R/S			3 highest post-treatment R/S		
		≦ 5	≦ 10	≦ 20	≦ 40	≦ 80	≦ 160	≦ 320						
1963	13	92	92	100					4	4	14	6	6	6
1964	17	59	100						6	9	10	6	7	7
1965	24	79	92	100					8	11	20	13	15*	27
1966	23	48	96	100					7	8	14	35*	52*	120*
1967	27	4	30	75	100				24	25	29	40*	41*	46
1968	32	3	53	91	100				22	23	24	31*	30*	42*
1969	17	0	18	65	100				26	26	27	48	52	46
1970	26	0	4	54	88	100			44	49	55	82	96	139
1971	14	0	0	0	7	57	86	100	150	270	305	238	272*	350

+On these farms a variety of OP-compounds have been used for fly control in 1963-71 including dimethoate for 2-8 seasons.

*Farms with commercial or private fly control outside the district for small scale trials.

REFERENCES

Abdel-Gawaad, A.A.W. and El-Gayar, F.H. (1972) Housefly control in Alexandria and its effect on the prevalence of certain summer diseases, *Z. Ang. Ent.*, 71, 44-52

Arevad, K. (1968-1969) Laboratory investigations on the behaviour of newly emerged houseflies, *Governt. Pest. Inf. Lab., Denmark, Ann. Report* 1967 (1968): 56-58, 1968 (1969): 54-55

Arevad, K. (1972) The orientation of newly emerged houseflies, *13 Int. Congr. Ent. Moscow* 1968, 3, 117

Ascher, K.R.S. (1961) Houseflies in Israel. III. Some observations at breeding sites in rural areas, and considerations about the influence of Israeli manure handling methods on housefly breeding, *Z. Ang. Ent.*, 48, 115-162

Austen, E.E. (1926) The Housefly.... *British Mus. (Nat. Hist.) Econ. Ser.*, No. 1a

Axtell, R.C. (1963a) Acarina occurring in animal manure, *Ann. Ent. Soc. Am.*, 56, 628-633

Axtell, R.C. (1963b) Manure inhabiting *Macrochelidae* predaceous on the housefly, *Adv. Acarology*, 1, 55-59

Axtell, R.C. (1963c) Effect of *Macrochelidae* on housefly production from dairy cattle manure, *J. econ. Ent.*, 56, 317-321

Axtell, R.C. (1968) Integrated housefly control: Populations of fly larvae and predaceous mites, *Macrocheles muscaedomesticae*, in poultry manure after larvicide treatment, *J. econ. Ent.*, 61, 245-249

Axtell, R.C. (1969) *Macrochelidae* as biological control agents for synanthropic flies, *Proc. 2nd Int. Congr. Acarol. Nottingham*, 401-416

Axtell, R.C. (1970a) Integrated fly-control program for caged-poultry houses, *J. econ. Ent.*, 63, 400-405

Axtell, R.C. (1970b) Fly control in caged-poultry houses: Comparison of larviciding and integrated control programs, *J. econ. Ent.*, 63, 1734-1737

Axtell, R.C. (1971) Ultralowvolume aerial sprays for the control of *Hippelates* gnats and other flies, *J. Georgia Entomol. Soc.*, 6, 101-105

Bailey, D.L., LaBrecque, G.C. and Bishop, P.M. (1967) Residual sprays for the control of houseflies, *Musca domestica,* in dairy barns, *Fla. Entomol.,* 50, 161-163

Bailey, D.L., LaBrecque, G.C. and Whitfield, T.L. (1970a) Insecticides in dry sugar baits for control of houseflies in Florida dairy barns, *J. econ. Ent.,* 63, 2000-2001

Bailey, D.L., LaBrecque, G.C. and Whitfield, T.L. (1970b) Resistance of houseflies (Diptera: Muscidae) to dimethoate and ronnel in Florida, *Fla. Entomol.,* 53, 1-5

Bailey, D.L., LaBrecque, G.C. and Whitfield, T.L. (1971) Resistance of houseflies (Diptera: Muscidae) in Florida to trichlorfon and dichlorovos formulated in sugar baits, *Can. Entomol.,* 103, 853-856

Balan, N.K. (1970) On the question of changes in the susceptibility of *Musca domestica* to chlorofos as a stomach poison, *Med. Parazit. (Mosk.),* 39, 491-497

Bath, S.S. and Stalker, J.M. (1970) A survey of Canadian populations of the housefly for resistance to insecticides, *J. econ. Ent.,* 63, 1947-1950

Bennett, F.D. and Legner, E.F. (1967) Memorandum on the possibilities of biological control of the muscid flies by hymenopterous parasites, WHO/VBC/67.22: 1-10 (Unpublished document)

Brady, U.E. Jr., Meifert, D.W. and LaBrecque, G.F. (1966) Residual sprays for the control of houseflies in field tests, *J. econ. Ent.,* 59, 1522-1523

Brown, A.W.A. and Pal, R. (1971) Insecticides resistance in arthropods, WHO Monograph Series No. 38, 2nd edit. (Geneva), 491 pp.

Bushland, R.C. (1971) Sterility principle for insect control. Historical development and recent innovations. Approaches to housefly control. In: Sterility principle for insect control or eradication: 8-11, Internat. Atom Energy Agency, Vienna, STI/PUB 1265, 542 pp.

Campion, D.G. (1972) Insect chemosterilants: A review, *Bull ent. Res.,* 61, 577-635

Carlson, D.A., Mayer, M.S., Silhacek, D.L., James, J.D., Morton Beroza, Bierl, B.A. (1971) Sex attractant pheromone of the housefly: Isolation, identification and synthesis, *Science,* 174, 76-78

C D C (1972) Public health pesticides, *Pest Control*, 40, 35-65

Davies, M.S. (1966) Field trials with synergized pyrethrins in the Todd insecticidal fog applicator (TIFA), *Pyrethrum Post*, 8(4), 18-22

Davies, M.S. (1971) An account of the development of bioallethrin and bioresmethrin synergized by piperonyl butoxide for control of *Musca domestica* L., *3rd Brit. Pest Control Conf.*, 3rd Session: Paper No. 6, 1-3

Dobson, R.C., Kurz, F.W. (1970) Control of houseflies in swine-finishing units by improved methods of waste disposal, *J. econ. Ent.*, 63, 171-174

Duport, M., Combiesco, L.and Enesco, A. (1968) Traitement experimental de l'espece *Musca domestica* par du thiotepa, sur le terrain, *Arch. roum. Path. exp.*, 27, 707-714

Duport, M., Combiesco, I. and Enescu, A. (1972) Traitement experimental de l'espece *Musca domestica* L. par du thiotepa sur le terrain, *13 Int. Congr. Ent., Moscow 1968*, 3, 419-421

Döhring, E. (1966) Zur Wirksamkeit und Anwendungstechnik von Fliegenstreifen, *Z. ang. Zool.*, 53, 95-121

Eastwood, R.E., Kada, J.M., Schoenburg, R.B. and Brydon, H.W. (1967) Investigations on fly control by composting poultry manures, *J. econ. Ent.* 60, 88-89

Enan, O.H. (1970) Susceptibility of housefly *Musca domestica* var. *vicina* to some insecticides in Alexandria area, U.A.R., *J. Egypt. Publ. Hlth. Ass.*, 45, 443-450

Erőss, J. and Mahunka, S. (1971) Investigations on coprophilous and stercoricolous macrochelids (Acari, Gamasina) in Hungary, as possible agent in the control of synanthropic flies, *Parasit. Hung.*, 4, 215-226

Foster, G.C., Whitten, M.J., Prout, T. and Gill, R. (1972) Chromosome rearrangements for the control of insect pests, *Science*, 176, 875-880

Funder, J.F. (1972) Fly control on farms with pyrethrum as the sole insecticide, *Danish Pest Inf. Lab. Ann. Report for 1971*, 32-33

Funder, J.F. (1973) Test of light traps for fly control on Danish farms, *Danish Pest Inf. Lab., Ann. Report for 1972*, 28-29

Georghiou, G.P. (1966) Distribution of insecticide-resistant houseflies on neighbouring farms, *J. econ. Ent.*, 59, 341-346

Georghiou, G.P. and Bowen, W.R. (1966) An analysis of housefly resistance to insecticides in California, *J. econ. Ent.*, 59, 204-214

Georghiou, G.P. and Hawley, M.K. (1971) Insecticide resistance resulting from sequential selection of houseflies in the field by organophosphorous compounds, *Bull. Wld Hlth Org.*, 45, 43-51

Georghiou, G.P., Hawley, M.K., Loomis, E.C. and Coombs, M.F. (1972) Insecticide resistance in houseflies in California, *Calif. Agric.*, Sept. 72, 4-6

Gotaas, H.B. (1956) Composting. Sanitary disposal and reclamation of organic waste, WHO Monograph Ser. No. 31, 205 pp.

Groth, U. and Berndt, K.P. (1970) On the behaviour of flies in animal quarters in relation to their control, *Angw. Parasitol.*, 11, 203-213

Gvozdeva, I.V., Artukhina, I.N., Storozheva, E.M. and Serafimova, A.M. (1972) Changes of susceptibility to organophosphorus insecticides after their many years' use in a number of populations of *Musca domestica* L. and *Cimex lectularius* L. from several regions of the USSR, *13 Internat. Congr. Entomol. Moscow 1968 Proceed.*, 3, 172

Hannine, S. ben, and J. Keiding (1967-1968) Strategic fly control in piggeries and cowsheds, *Govt. Pest Inf. Lab., Denmark, Ann. Report 1966*, 35-37 (1967) *Ann. Report 1967*, 37-41 (1968)

Hansens, E.J., Benezet, H.J. and Evans, E.S. Jr. (1967) Housefly control and insecticide resistance with continued use of diazinon, ronnel, and dimethoate, *J. econ. Ent.*, 60, 1057-1064

Hansens, E.J., Evans, E.S. and Shibles, D.B. (1968) Tests of bromophos for housefly control in New Jersey, *J. econ. Ent.*, 61, 833-836

Hansens, E.J. and Anderson, W.F. (1970) Housefly control and insecticide resistance in New Jersey, *J. econ. Ent.*, 63, 1924-1926

Hayashi, A., Mukai, S. and Matsuzaki, S. (1972) The resistance level of the housefly to several synthetic insecticides in Kochi prefecture, Japan, *Botyu-Kagaku*, 37, 91-93

Hecht, O. (1970) Light and color reactions of *Musca domestica* under different conditions, *Bull. Ent. Soc. Am.*, 16, 74-78

Holway, R.T., Mitchell, W.A. and Salah, A.A. (1951) Studies on the seasonal prevalence and dispersal of the Egyptian housefly. II. The larvae and their breeding areas, *Ann. Ent. Soc. Am.*, 44, 489-510

Johnson, W.H. (1959) Sanitation in the control of insects and rodents of public health importance. Communicable Disease Center, U.S. Public Health Service. Training Guide. Rodent Control Series, IV, 1-42

Keiding, J. (1956) Resistance to organic phosphorus insecticides of the housefly, *Science*, 123, 1173-1174

Keiding, J. (1964) Control of houseflies, WHO Inf. Circ. Vector Control, 3, 19-35, (Unpublished document)

Keiding, J.(1965a) Observations on the behaviour of the housefly in relation to its control, *Riv. Parasit.*, 26, 45-60

Keiding, J. (1965b) Development of resistance in field populations of houseflies exposed to residual treatments with organophosphorus compounds, *Meded. Landb. Hogesch. Gent.*, 30, 1362-1381

Keiding, J. (1967) Persistence of resistant populations after the relaxation of the selection pressure, *World Rev. Pest Control*, 6, 115-130

Keiding, J. (1968) Insecticide resistance in houseflies, *Govt. Pest Inf. Lab., Denmark, Ann. Report, 1967*, 46-54

Keiding, J. (1969) Insecticide resistance in houseflies, *Govt. Pest Inf. Lab., Denmark,Ann. Report, 1968*, 39-44

Keiding, J. (1970) Insecticide resistance in houseflies, *Danish Pest Inf. Lab. Ann. Report, 1969*, 43-50

Keiding, J. (1972) Insecticide resistance in houseflies, *Danish Pest Inf. Lab. Ann. Report, 1971*, 37-44

Keiding, J. and Hannine, S. ben (1967) Fly behaviour and distribution in piggeries and cowsheds, *Govt. Pest Inf. Lab., Denmark, Ann. Report, 1966*, 53-54

Keiding, J. and Hannine, S. ben (1972) Selective housefly control on farms based on the resting habits of the flies, *13 Internat. Congr. Entomol., Moscow 1968*, 3, 186

Kerdpibule, V. and Hirakoso, S. (1971) Susceptibilities to several insecticides in adult houseflies collected from various districts in Thailand, *Japan J. Exp. Med.*, 41, 541-545

Kilpatrick, J.W. and Quarterman, K.D. (1952) Field studies on the resting habits of flies in relation to chemical control. II. In Rural areas, *Am. J. Trop. Med. Hyg.*, 1, 1026-1031

Kilpatrick, J.W. and Schoof, H.F. (1956) The use of insecticide treated cords for housefly control, *Publ. Health Reports* (U.S. Dept. Hlth. etc.) 71, 144-150

Kilpatrick, J.W. and Schoof, H.F. (1963) Effectiveness of seven organophosphorus compounds as space applications against *Musca domestica*, *J. econ. Ent.*, 56, 560-563

LaBrecque, G.C. and Meifert, D.W. (1966) Control of houseflies in poultry houses with chemosterilants, *J. med. Ent.*, 3, 323-326

LaBrecque, G.C. and Weidhaas, D.E. (1970) Advances of integrating sterile-male releases with other methods of control against houseflies, *J. econ. Ent.*, 63, 379-382

LaBrecque, G.C., Meifert, D.W. and Weidhaas, D.E. (1972) Dynamics of housefly and stable fly populations, *Fla. Ent.*, 55, 101-106

Lancaster, J.L. and Simco, J.S. (1969) Housefly control by chemical sterilization with apholate, *Bull. Agric. Exp. Station, Arkansas Univ.*, 737, 12 pp.

Legner, E.F. and Brydon, H.W. (1966) Suppression of dung-inhabiting fly populations by pupal parasites, *Ann. Ent. Soc. Amer.*, 59, 638-651

Legner, E.F. and McCoy, C.W. (1966) The housefly, *Musca domestica* L as an exotic species in the western hemisphere incites biological control studies, *Can. Entom.*, 98, 243-248

Legner, E.F., Bay, E.C. and White, E.B. (1967) Activity of parasites from Diptera: *Musca domestica*, *Stomoxys calcitrans*, *Fannia canicularis*, and *F. femoralis*, at sites in the western hemisphere, *Ann. Ent. Soc. Amer.*, 60, 462-468

Legner, E.F. and Olton, G.S. (1968a) Activity of parasites from Diptera: *Musca domestica*, *Stomoxys calcitrans*, species of *Fannia muscina* and *Ophyra*. II. At sites in the Eastern Hemisphere and Pacific area, *Ann. Ent. Soc. Am.*, 61, 1306-1314

Legner, E.F. and Olton, G.S. (1968b) The biological method and integrated control of house and stable flies in California, *California Agriculture*, June 68, 2-4

Legner, E.F. and Greathead, D.J. (1969) Parasitism of pupae in East African populations of *Musca domestica* and *Stomyxys calcitrans*, *Ann. Ent. Soc. Amer.*, 62, 128-133

Legner, E.F. and Olton, G.S. (1970) Worldwide survey and comparison of adult predator and scavenger insect populations associated with domestic animal manure where livestock is artificially congregated, *Hilgardia*, 40, 225-266

Legner, E.F. (1971) Some effects of the ambient arthropod complex on the density and potential parasitization of Muscoid Diptera in poultry wastes, *J. econ. Ent.*, 64, 111-115

Lewis, J.B. and Sawicki, R.M. (1971) Characterization of the resistance mechanisms to diazinon, parathion and diazoxon in the organophosphorus-resistant SKA Strain of houseflies (*Musca domestica* L.), *Pesticide Bioch. and Physiol.*, 1, 275-285

Lineva, V.A. (1966) The pattern of the effect of chlorophos upon the ovogenesis of houseflies (*Musca domestica* L.) in its long application under natural conditions, *Med. Parasit.(Mosk.)* 35, 217-223

Lineva, V.A. (1970) The influence of the prolonged action of khlorofos on a natural population of houseflies (*Musca domestica* L.), *Med. Parasit. (Mosk.)*, 39, 73-77

Lineva, V.A. (1972) Prolonged exposure of neutral populations of housefly (*Musca domestica* L.) to chlorophos, *13 Internat. Congr. Ent., Moscow 1968, Proceed.*, 3, 197-198

Loaeza, R. McG. and Corona, A.O. (1965) Esterilization de la *Musca domestica* on apholate, *Folia Entomol. Mexicana.*, Sept. 1965, 3-15

Lofgren, C.S., Ford, H.R., Tonn, R.J. and Jatanasen, S. (1970) The effectiveness of ultra-low-volume application of malathion at a rate of 5 fluid ounces per acre in controlling *Aedes aegypti* in a large-scale test at Nakhon Sawan, Thailand, *Bull. Wld Hlth Org.*, 42, 15-25

Lindquist, A.W. (1956) Effectiveness of organophosphorus insecticides against insects, *Bull. Wld Hlth Org.*, 16, 33-39

Magaudda, P.L., Sacca, G. and Guarniera,D. (1969) Sterile male method integrated by insecticides for the control of *Musca domestica* L. in the island of Vulcano, Italy, *Ann. Ist. Super. Sanita (Rome)*, 5, 29-38

Maier, P.P., Baker, W.C., Bogue, M.D., Kilpatrick, J.W. and Quarterman, K.D. (1952) Field studies on the resting habits of flies in relation to chemical control. Part I. In urban areas, *Am. J. Trop. Med. Hyg.*, 1, 1020-1025

Mathis, W. and Schoof, H.F. (1968) Chemical control of houseflies in dairy barns and chicken ranches, *J. econ. Ent.*, 61, 1071-1073

Mathis, W., Flynn, A.D., Schoof, H.F. (1972) Housefly resistance in the Savannah, Georgia area and its influence on chemical-control measures, *J. econ. Ent.*, 65, 748-751

Matsuzawa, H. and Fuji'i, Y. (1968) On the sterilizing effect of Hempa on the common housefly, *Musca domestica vicina*. II. A field test in Sei Island, *Jap. J. San. Zool.*, 19, 210-212

Meifert, D.W., LaBrecque,G .C., Smith, C.N. and Morgan, P.B. (1967) Control of houseflies on some West Indies islands with metepa, apholate, and trichlorfon baits, *J. econ. Ent.*, 60, 480-485

Meifert, D.W., LaBrecque, G.C. and Rye, J.R. Jr. (1969) Housefly, *Musca domestica*, control with chemosterilants and insecticides, *Florida Entomol.*, 52, 55-60

Meifert, D.W. and LaBrecque, G.C. (1971) Integrated control for the suppression of a population of houseflies, *Musca domestica* L., *J. med. Ent.*, 8, 43-45

Mer, G.G. and Paz, M. (1960) Control of houseflies in rural areas by means of traps, *Riv. Parasit.*, 21, 143-150

Millar, L.A. and Hooper, C.H.S. (1969) Resistance to insecticides of three strains of *Musca domestica* L. from Brisbane, Queensland, *J. Austral. Ent. Soc.*, 8, 113-115

Miller, R.W. (1970) Larvicides for fly control - a review, *Bull. Ent. Soc. Amer.*, 16, 154-158

Miller, R.W. and Gordon, C.H. (1972a) Encapsulated Rabon for larval housefly control in cow manure, *J. econ. Ent.*, 65, 455-458

Miller, R.W. and Gordon, C.H. (1972b) Technical Rabon for larval housefly control in cow manure, *J. econ. Ent.*, 65, 1064-1066

Morgan, N.O., Pickens, L.G. and Thimijan, R.W. (1970) Houseflies and stable flies captured by two types of traps, *J. econ. Ent.*, 63, 672-673

Mourier, H. (1972) Release of native pupal parasitoids of houseflies on Danish farms, *Vidensk. Meddr. dansk Naturh. Fore*, 135, 129-137

Mourier, H. (1973) Field trials on release of sterile houseflies on Danish farms, *Danish. Pest Inf. Lab. Ann. Report, 1972* (in press)

Mulder, R. (1972) 1-(2,6-disubstituted-benzoyl)-3-(phenyl)-ureas, a new group of insecticidal compounds, *14 Internat. Congr. Entomol., Canberra 1972*, Abstracts, 259

Neselovskaya, V.K., Goldina, G.S. and Ryazantsev, V.F. (1968) The use of khlorofos fly-killers in combination with food baits for the control of synanthropic flies, *Med. Parazit. (Mosk)*, 37, 230-232

Omori, N. and Suenaga, O. (1962) A method of destructing the maggots of housefly and stablefly animal manure by airtight vinyl cover, *Endem. Disease Bull. of Nagasaki Univ.* 4(1), 52-56

Oppenoorth, F.J. (1971) Resistance in insects: The role of metabolism and the possible use of synergists, *Bull. Wld Hlth Org.*, 44, 195-202

Patton, W.S. (1931) Insect, ticks, mites and venomous animals of medical and veterinary importance, II. Public Health, Croydon, England, 740 pp.

Pausch, R.D. (1972) Local housefly control with baited chemosterilants. 2: Field application in enclosed situations, *J. econ. Ent.*, 65, 449-450

Peck, J.H. (1969) Arthropod predators of immature Diptera developing in poultry droppings in Northern California. Part II. Laboratory studies on feeding behaviour and predation potential of selected species, *J. med. Ent.*, 6, 168-171

Peck, J.H. and Anderson, J.R. (1969) Arthropod predators of immature Diptera developing in poultry droppings in Northern California. Part I. Determination, seasonal abundance and natural cohabitation with prey, *J. med. Ent.*, 6, 163-167

Peck, J.H. and Anderson, J.R. (1970) Influence of poultry-manure-removal schedules on various Diptera larvae and selected arthropod predators, *J. econ. Ent.*, 63, 82-90

Pickens, L.G., Morgan, N.O. and Thimijan, R.W. (1969) Housefly response to fluorescent lamps: Influenced by fly age and nutrition, air temperature, and position of lamps, *J. econ. Ent.*, 62, 536-539

Pimentel, D. (1954) Relationship of ants to fly control in Puerto Rico, *J. Trop. Med. Hyg.*, 4, 147-151

Pimentel, D. and Uhler, L. (1969) Ants and the control of house flies in the Philippines, *J. econ. Ent.*, 62, 248

Privora, M., Pinterova, J. and Maskova, L. (1969) Meat proteins as an attractant for the housefly, *Congr. Int. Antiparasit. Milan 1969*, 33, 1-6

Reid, J.A. (1953) Notes on houseflies and blowflies in Malaya, *Bull. Inst. Med. Res. Fed. Malaya*, 7, 1-26

Robbins, W.E., Thompson, M.J., Yamamoto, R.T. and Shortino, T.J. (1965) Feeding stimulants for the female housefly *Musca domestica* L.,*Science*, 147, 628-630

Rodrigues, J.G. (1972) The manure acarine complex and its significance in muscid control, *13 Int. Congr. Entomolog. Moscow, 1968*, 3, 236

Ross Institute (1968) Rural sanitation in the tropics, Bull. No. 8, (Ross Inst. Inform. and Advisory Service) London, 47 pp.

Roubaud, E. (1936) The biothermic method of fly destruction and the ease with which it can be adapted to rural conditions, *Quart. Bull. Hlth. Org. League Nat.*, 5, 214-218

Sacca, G., Scirocchi, A., De Meo, G.M. and Mastrilli, M.L. (1966) Una prova di campo con il chemosterilante Hempa contro *Musca domestica* L., *Atti Soc. Peloritana Sci. fis. matem. natur.*, 12, 1-2

Sawicki, R.M. and Farnham, A.W. (1968) Examination of the isolated autosomes of the SKA strain of houseflies for resistance to several insecticides with and without pretreatment with sesamex and TBTP, *Bull. ent. Res.*, 59, 409-421

Sawicki, R.M. (1970a) Interaction between factors of resistance in a strain of houseflies polygenic for resistance to organophosphorus insecticides, *7 Internat. Plant Prot. Congr. Paris, 1970*, 808-811

Sawicki, R.M. (1970b) Interaction between the factor delaying penetration of insecticides and the desethylation mechanism of resistance in organophosphorus-resistance houseflies, *Pest. Sci.*, 1, 84-87

Sawicki, R.M. (1972) Genetics of resistance of houseflies to dimethoate, Rothamsted Report for 1971, 179-180

Sehgal, B.S. and Kumar, P. (1970) Seasonal fly control through baiting in rural areas, *Indian J. Publ. Hlth.*, 14, 116-120

Sento, T. (1961) Fly control experiments by improved treatment of animal manure, *Endem. Dis. Bull. of Nagasaki Univ.*, 3(2), 139-144

Sento, T. (1965) Fly control experiment by the large four roomed closed tank for animal manure, *Endem. Dis. Bull. of Nagasaki Univ.*, 7(1), 38-63

Singh, P. and Rodrigues, J.G. (1969) *Macrocheles muscaedomesticae*, current knowledge and status in biological control, WHO/VBC/ 69.162, 1-9, (Unpublished document)

Smith, R.H. and von Borstel, R.C. (1972) Genetic control of insect populations, *Science*, 178, 1164-1174

Sztankay-Gulyas, M. and Eröss, J. (1970) The susceptibility of *Musca domestica* to new contact poisons in Hungary, *Magya Allatorvosok Lapja*, 25, 357-360

Tarry, D.W., Kirkwood, A.C. and Hebert, C.N. (1971) The response to "black-light" radiation of some common flies of economic importance, *Ent. Exp. Appl.*, 14, 23-29

Thimijan, R.W., Pickens, L.G. and Morgan, N.O. (1970), A trap for houseflies, *J. econ. Ent.*, 63, 1030-1032

Thimijan, R.W., Pickens, L.G., Morgan, N.O. and Miller, R.W. (1972) Housefly capture as a function of number of traps in a dairy barn, *J. econ. Ent.*, 65, 876-877

Thomsen, M. (1934) Fly control in Denmark, *Quart. Bull. Hlth Org. League of Nations*, 3, 304-324

Toppozada, A., Mohamed, M.A. and Eldefrawi, M.E. (1967) Resistance of the housefly (*Musca domestica* L.) to insecticides in Egypt, *Bull. Wld Hlth Org.*, 36, 937-948

Treece, R.E. and Molineros, J.A. (1971) Fly control in dairy barns (Dipt.), *Ohio Rep. Res. and Develop.*, 56, 30

Treece, R.E. and Molineros, J.A. (1972) Fly control in Ohio barns... II, *Ohio Agric. Res. and Dev. Center. Res. Circ.*, 189, 1-8

Varma, R.N., Mahadevan, B. and Dev Chaudhary, S.S.K. (1966) Housefly control in dairy farms with malathion impregnated cords. Part 1., *Armed Forces Medical J. India*, 22, 281-296

Varma, R.N., Mahadevan, B. and Lopez, W.A. (1967) Housefly control in dairy farms with malathion impregnated cords. Part II., *Armed Forces Medical J. India*, 23, 123-126

Wagner, E.G. and Lanoix, J.N. (1958) Excreta disposal for rural areas and small communities, WHO Monograph Ser. No. 39, 187 pp.

Wagoner, D.E., Johnson, O.A. and Nickel, C.A. (1971) Fertility reduced in a caged native housefly strain by the introduction of strains bearing heterozygous chromosomal translocations, *Nature*, 234, 473-475

Walsh, J.D. (1964) A survey of fly production in cattle feedlots in the San Joaquin Valley, *Calif. Vector Views*, 11, 33-39

Weidhaas, D.E. (1968) Field development and evaluation of chemosterilants, In: Principles of insect chemosterilization, North-Holl. Publ. Co. Amsterdam. edit. LaBrecque, G.C. *et al.* 354 pp., pp. 275-314

Weidhaas, D.E. and LaBrecque, G.C. (1970) Studies on the population dynamics of the housefly, *Musca domestica*, *Bull. Wld Hlth Org.*, 43, 721-725

West, L.S. (1951) The housefly, Comstock Pub. Co. Ithaca, N.Y., 584 pp.

Wharton, R.H., Seow, C.L., Ganapathipillai, A. and Jabaratnam, G. (1962) Housefly populations and their dispersion in Malaya with particular reference to the fly problem in the Cameron Highlands, *Med. J. Malaya*, 17, 115-131

Whitten, M.J. (1970) Genetics of pests in their management, In: Concepts of pest management, R.L. Rabb and F.E. Guthrie (ed.) Raleigh, USA, 119-135

Whitten, M.J. (1971) Insect control by genetic manipulation of natural populations, *Science*, 171, 682-684

WHO (1970) Insecticide resistance and vector control, 17th Rep. WHO Expert Committee on Insecticides (Nov. 1968), *Wld Hlth Org. techn. Rep. Ser.*, 443, 279 pp., 196-205

Wichmand, H. (1953) Control of multi-resistant houseflies (with parathion bands), *Nature*, 172, 758-759

Wiesmann, R. (1960) Untersuchungen über die Sinnesfunktionen der Antennen von *Musca domestica* L. in Zusammenhang mit dem Köderproblem, *Mitt. Schweiz. Ent. Ges.*, 33, 121-154

Wicht, M.C. and Rodrigues, J.G. (1970) Integrated control of Muscid flies in poultry houses using predator mites, selected pesticides and microbial agents, *J. med. Ent.*, 7, 687-692

Wilkinson, C.F. (1968) The role of insecticide synergists in resistance problems, *Wld. Rev. Pest Contr.*, 7, 155-168

Wilkinson, C.F. (1971) Effects of synergists on the metabolism and toxicity of anticholinesterases, *Bull. Wld Hlth Org.*, 44, 171-190

Wright, J.E. and Spates, G.E. (1972) Laboratory evaluation of compounds to determine juvenile hormone activity against the stable fly, *J. econ. Ent.*, 65, 1346-1349

Yasutomi, K. and Keiding, J. (1969) Effect of synergists on resistance to insecticides (in houseflies), *Govern. Pest Inf. Lab. Denmark, Ann. Report 1968*, 45-46

Zakharova, N.F. (1971) Experiments on hempa as a chemosterilant of houseflies, *Med. Parazit. (Mosk.)*, 40, 21-28

Note: For additional references see page 34

DISCUSSION OF THE PAPER ON HOUSEFLIES,

MUSCA DOMESTICA

Elton J. Hansens

Rutgers, The State University

New Brunswick, New Jersey, U.S.A.

After reading Mr Keiding's paper my first reaction was "What more is there to say?". On further reflection there are some experiences and ideas that come to mind from our work in New Jersey over the past 25 years which might be worth considering.

With the advent of DDT, farmers for the first time were able to control flies in barns for an entire season with a single spray application. However, by 1947 and 1948 DDT was no longer giving control at many locations and subsequently DDT was banned in dairy barns because of the possibilities of accumulation in milk. Within two to three years, the substitutes methoxychlor and lindane had failed, and use of dieldrin on manure for larval control resulted in high resistance within a single season. We then recognized that a residual against larvae was a more rapid method of building resistance, since both male and female flies were selected before mating was possible. Before the chlorinated hydrocarbons failed completely, diazinon proved a very effective organophosphorus insecticide. Subsequently we also recommended malathion, ronnel and dimethoate. However in each case after two to three years,we were faced with resistance. Ronnel seemed to give a slower build-up of resistance than the other OP compounds contrary to the experience in Denmark. OP resistance soon was widespread and laboratory studies showed resistance to OP compounds greatly enhanced the level of resistance to the chlorinated hydrocarbons. This increase, Forgash et al. (1962) showed, could be partly accounted for by poorer penetration of diazinon into resistant strains.

This history corresponds quite closely with the experience in Denmark. In New Jersey we now find that no residual will give control in some areas when used throughout a season, while good control can be achieved in other areas. Whether residual materials are effective or not depends on the magnitude of the fly population and the history of insecticide use. These are just other words for Mr Keiding's "high breeding potential and continuous insecticide pressure".

When we look to the future we must first admit that we now have more insecticides and more methods of attacking the fly problem than ever before in the history of mankind. At the same time, we know the housefly has a high degree of adaptability when we introduce control measures. We then have the problem of careful selection and careful application of control methods.

Sanitation should always be the watchword in fly control and removal of all possible breeding material is the preferred approach for control of the housefly. However, farmers often neglect sanitation, particularly when they are busy during the growing season with a multitude of other tasks. Location of all breeding places is virtually impossible, and even with the best sanitation insecticidal control is usually necessary.

Residual treatments where they can be used, furnish us with the simplest long-range control. When the residual fails, baits or space sprays are the next line of defense. In New Jersey a combination of one early season residual treatment followed by the use of baits and/or space sprays has been satisfactory, with some exceptions.

Our experience with resistance has convinced us that the housefly is one of the most adaptable of insects, able to find and successfully invade a wide range of ecological situations, and to alter its biology and physiology readily to exploit the changing environment. This we have seen it do most dramatically with insecticide resistance. Even chemosterilants may not be a complete barrier against the housefly. Abasa and Hansens (1969) over a period of generations selected a strain which was able to persist when fed sugar with 10 times the amount of apholate that gave sterility in the early generations. Only a few individuals which possess genes providing an alternate pathway around a chemical can survive and are able to produce a strain which is successful in the face of the new hazard.

I would agree with Mr Keiding that one of the great needs in housefly control would be for a really effective attractant; an attractant which would draw houseflies even in the presence of adequate food supply and breeding conditions. If we had such an attractant then it would be possible to use a wide variety of toxicants or chemosterilants with minimum hazard to the environment.

In dealing with resistant strains we should keep in mind that the selective use of an insecticide selects not only for resistance but also for other characters in the gene pool. We have reared a dimethoate resistant strain whose life cycle was considerably shorter than normal. On the other hand, most resistant strains have slightly longer cycles.

We have noted differences between susceptible and resistant strains in regard to temperature effects and with regard to moisture effects. The resistant A-Strain tended to pupate in drier areas than the susceptible strains. These subtle biological differences have great possibilities for development of new control measures. If resistant strains prefer drier areas, they should then be more subject to parasites and predators. If the life cycle is lengthened, then in the course of a season we could expect one or two less generations with a very large attendant decrease in the total population for the year. On the other hand, the problem would become compounded by a life cycle which is shorter. Not only must we look for new materials but for completely new approaches which could effectively exploit the biology of the housefly to its own disadvantage. Mr Keiding's detailed studies of behaviour emerging flies are the kind of research I mean.

There is rather inadequate information on mating habits, on dispersal, on population densities, and on over-wintering. Circumstantial evidence from studies of resistance would lead us to believe that flies in a local area are capable of surviving through the winter, whether as adults or larvae we do not really know. Inside some buildings,flies are able to continue their cycle throughout the year. Decline of the resistance levels in the spring in areas where resistance was high the previous season indicates susceptible flies have moved into the treated area. Individual flies will easily move from one to five miles and we have observed a fly marked with a coloured dye which was captured a mile from the release point within an hour. We need to watch for changed biology as new strains appear and alter present controls or invent new ones accordingly.

Whether residual insecticides, chemosterilants, juvenile hormones, baits, or space sprays are to be used, the limitation of chemicals in favour of other means of control seems desirable, both because of environmental contamination and because of resistance to pesticides. Reduced insecticide use by fewer applications or by better selection of areas to be treated would enable better utilization of parasites and predators. An excellent attractant would be very helpful. Until new sophisticated and imaginative controls are developed we do have a large arsenal of chemicals and approaches. Of all the possibilities, source reduction is the best known and most feasible method for eliminating the flies whether the problem is in an area of high resistance or none at all.

I would close with a question and a comment. The question - "Is it necessary to control the housefly at all in dairy barns?". The comment - In New Jersey dairy barns, houseflies are no longer a hazard in disease transmission. Cows are milked by machine, and the milk flows directly into cooling tanks in the milkhouse. Milk, from the time it is produced in the cow until it reaches the processing plant, is never exposed to the open air or to contamination by flies. Fly control is then a matter of comfort to the farmer and the animals.

REFERENCES

Abasa, R.O. and Hansens, E.J. (1969) An apholate-resistance strain of houseflies. I. Resistance to other chemosterilants and to insecticides, *J. econ. Ent.*, 62, 334-338

Forgash, A.J., Cook, B.J. and Riley, R.C. (1962) Mechanisms of resistance in diazinon-selected multi-resistant *Musca domestica*, *J. econ. Ent.*, 55, 544-551

ADDITIONAL REFERENCES ON HOUSEFLIES FROM PAGE 30

Anderson, J.R. and Poorbaugh, J.H. (1964) Observations on the ethology and ecology of various Diptera associated with northern California poultry ranches, *J. med. Ent.*, 1, 131-147

Carlson, D.A. and Beroza, M. (1973) Field evaluations of (Z)-9-tricosene, a sex attractant pheromone of the housefly, *Env. Entomol.*, 2, 555-559

Davies, M., Keiding, J. and von Hofsten, C.G. (1958) Resistance to pyrethrins and to pyrethrins-piperonyl butoxide in a wild strain of *Musca domestica* L. in Sweden, *Nature*, 182, 1816-1817

El-Deeb, A.L., Hammad, S.M. and Gaaboub, I.A. (1969) Laboratory tests to determine susceptibility of adult females of the housefly to malathion, *Bull. Entomol. Soc. Egypt.*, *Econ. Ser.* 3, 179-185

Hayashi, A., Hatsukade, M., Hasegawa, M. and Hattori, K. (1971) Insecticide resistance of adult housefly in Hokkaido, *Jap. J. San. Zool.*, 22, 161-165

Sacca, G., Stella, E., Gandolfo, D. and Mastrilli, M.L. (1970) Prime osservazioni sui livelli di resistenza in *Musca domestica* L. nella provincia di Latina, a 24 anni dall'introduzinne degli contatto, *Parassitologia*, 12, 63-68

Volcsik, R. (1971) Die Wirkung von Attraktivstoffen auf synanthrope Fliegen, *Angew. Parasitol.*, 12, 96-109

TICKS WITH SPECIAL EMPHASIS ON *BOOPHILUS MICROPLUS*

R.H. Wharton

Division of Entomology, Long Pocket Laboratories

CSIRO, Indooroopilly, Queensland, Australia 4068

INTRODUCTION

Man, with his domestic animals, has intruded into many complex host-parasite relationships involving ticks. Indeed, with the exception of the form of *Ornithodorus porcinus* which feeds almost exclusively on man in the Kenya Highlands (Walton 1962), all of man's and many of his domestic animals' contacts with ticks are casual in that wildlife are the preferred hosts and the source of the disease organisms which ticks transmit. This interaction creates difficult control problems since it means that the habitat and the wild life hosts must be manipulated if control of the arthropod is to be achieved. In these situations, there are obvious difficulties which are technical, cultural and emotional. Area control by residual insecticides, or any insecticide, will become increasingly unpopular, and the destruction of wildlife will be equally distasteful to a world quite rightly concerned with environmental conservation. The following examples illustrate the problems: (1) In the Ozark region of Oklahoma, U.S.A., recreational areas are flooded with the lone star tick *Amblyomma americanum* which feeds very readily on man and cattle and many other hosts including the white-tailed deer *Odocoileus virginianus* (Hair and Howell 1970). Control of this tick will ultimately involve the control of the deer or its habitat, yet deer protection at this stage has led to such vast populations of the lone star tick that the tick itself has become a major mechanism in the regulation of deer populations (Bolte, Hair and Fletcher, 1970). (2) In Africa, cattle are constantly under challenge from several species of ticks which were originally, and still are, parasites of game animals. Permanent control of these ticks is obviously impossible without the elimination of the game animals including the buffalo

(Syncerus caffer), which may also be the source of perhaps the most menacing of all tick-transmitted diseases, East Coast Fever, caused by *Theileria parva* (Brocklesby 1966). For the present, there would appear to be only two practical approaches to problems such as these - for man, personal protection through repellents, and for cattle, the regular and frequent application of acaricides.

However, the most immediate and challenging tick problem is that presented by a natural tick-host relationship. This is the problem of *Boophilus microplus* which is a parasite of cattle and the vector of their *Babesia* blood parasites. *B. microplus* has been transported from Asia to all major countries lying within the tropical and sub-tropical regions of the world (Figure 1). It occurs throughout Asia extending north into China, Japan and Korea, in southern, central and northern America as far north as Mexico where it co-exists with *B. annulatus*, in the islands of the Caribbean, parts of eastern and southern Africa where it co-exists with another *Boophilus* species *B. decoloratus* and in New Guinea and Australia, but not New Zealand (Hoogstraal 1956; Roberts 1965).

In theory, *Boophilus* ticks should be relatively easy to control as they complete their parasitic life on one host, they feed predominantly on cattle, and, apart from horses and deer, other hosts can usually be ignored. In practice they are regarded as of minor importance in Africa, and they have been eradicated from areas in other countries; thus *B. annulatus* was eradicated from vast regions of the U.S.A. with a relatively inefficient acaricide, arsenic; *B. microplus* was also eradicated from Florida and Puerto Rico but with greater difficulty probably because of the presence of white-tailed deer (Anon., 1962). *B. microplus* was eradicated from areas in Papua New Guinea not inhabited by Rusa deer *Cervus timorensis* (Anderson 1963) and from limited areas towards its southern range in Argentina (L. Collado, personal communications) and in New South Wales, Australia (Mackerras et al., 1961). In recent times, an eradication campaign has been mounted against *B. microplus* on the island of St. Croix of the U.S. Virgin Islands (Anon., 1972). In other areas, and in Australia in particular, *B. microplus* continues to present a serious and, in some places, an increasingly serious problem.

CURRENT STATUS OF THE *BOOPHILUS MICROPLUS* PROBLEM

Control of *B. microplus* is necessary for two main reasons: (1) prevention of live body weight losses from the direct effects of ticks, and (2) prevention of clinical disease in cattle from infection with *Babesia* parasites transmitted by the tick. In most areas, enzootic for *B. microplus* tick fever transmission is maintained at a level where calves, born by immune mothers, become infected before weaning and show no clinical disease throughout

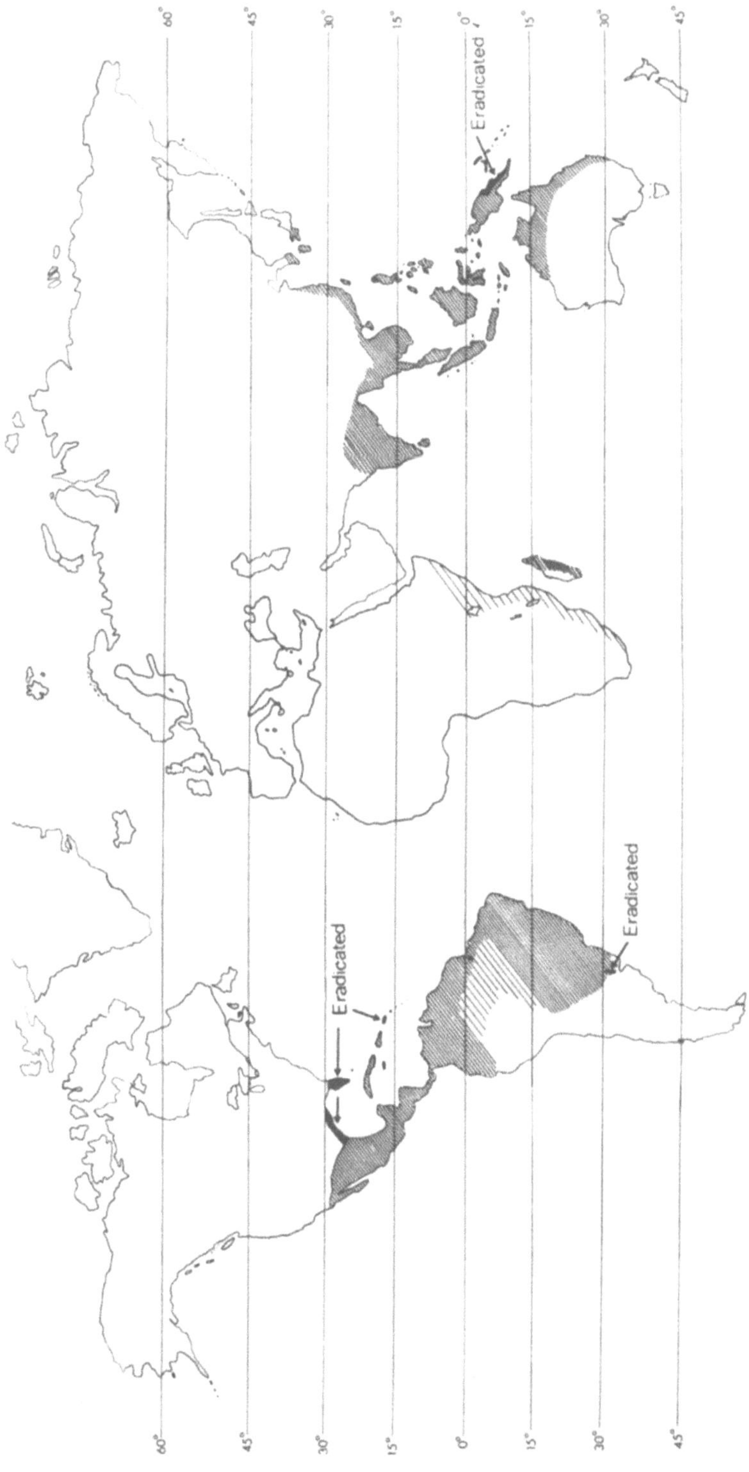

Figure 1. Distribution of *Boophilus Microplus*

their lives (Reik 1965). In areas marginal for tick survival or where control has reduced tick numbers to a level where tick fever transmission is interrupted, cattle may be protected by vaccination (Mahoney 1972). Similarly, non-immune cattle imported into an enzootic area can be protected by vaccination.

In Asia, Zebu (*Bos indicus*) cattle appear to co-exist satisfactorily with *B. microplus* and *Babesia* parasites but the introduction of exotic European (*Bos taurus*) cattle (mainly for milk production) has led to problems of tick control and tick fever (Legg 1959). At the present time, they are relatively minor. Although acaricide resistance to benzene hexachloride (BHC) has been reported from Malaya, Sarawak, India and Malagasy (Wharton and Roulston, 1970), alternative acaricides are readily available. However, unless efforts are made to evolve milking breeds of *Bos indicus x Bos taurus* cattle which are adapted to local conditions, problems could increase in the years ahead.

In Africa, *B. microplus* is of little significance at present, but it is feared that it may be extending its range in South Africa (C.J. Howell, personal communication). Whereas the acaricide resistance problem with *B. decoloratus* is not yet acute (J.F.A. Baker, personal communication), resistance in this species has developed to arsenic, DDT, BHC and related chemicals and to a limited number of organophosphorus chemicals (Wharton and Roulston, 1970).

In the Americas and in Australia, *B. microplus* is a major problem. It was introduced to both regions and spread to climatically favourable areas, e.g. in Australia to areas which are not too dry (500 mm rain) or too cold (more than four months with mean temperatures of 16°C) (Wilkinson 1970; Mackerras et al., 1961). Where the cattle industry is based on breeds of European (*Bos taurus*) origin, cattle owners have found it necessary to treat their animals with acaricides in order to prevent losses in productivity. This was the case in virtually all areas occupied by the tick in Australia and Argentina until recent years. On the other hand in Brazil, where the beef industry developed largely with cattle of Indian Zebu (*Bos indicus*) origin, tick control is practised only among dairy herds adjacent to major cities and in the southern state of Rio Grande de Sul where Hereford and Devon beef breeds predominate (N.K. Amaral, personal communication). Little information is available from other countries in southern and central America but criollo cattle, which are probably the dominant breed, have been shown to be more resistant to *B. microplus* than European cattle (Ulloa and de Alba, 1957) and thus help to reduce the severity of the *B. microplus* problem. In these countries and in Asia, where peasant farmers pen their cattle near or in houses at night, domestic fowls undoubtedly assist in limiting tick numbers. They have been shown to be tick predators (Theiler

1958),and the fact that most engorged females fall from cattle in the early morning (Wharton and Utech, 1970) would make them attractive and easy prey for undernourished hens around houses.

CONTROL BY ACARICIDES

Control based on the application of acaricides at frequent intervals to prevent *B. microplus* populations from increasing to damaging levels on herds of European (*Bos taurus*) cattle has led to crises, particularly in Argentina, Brazil and Australia. Resistance has developed in succession to arsenic, to chlorinated hydrocarbons and to organophosphorus (OP) and carbamate compounds. In most countries, because of resistance and/or because of unacceptable residues in meat, arsenic and the chlorinated hydrocarbons are no longer used.

Resistance to OP-acaricides in Australia

A chronological record of OP-resistance in relation to the introduction of OP-acaricides in Australia (Table 1) indicates the frightening rapidity with which *B. microplus* has adapted to the selection pressures of new acaricides. It is tragic to record that a high level of resistance was demonstrated in the field seven years after the introduction of the first OP-compound and that, less than three years after the alternative acaricides Dursban and bromophos ethyl were introduced, *B. microplus* had again responded to the challenge.

There are five, probably more, toxicologically and biochemically distinct types of resistance occurring on cattle in Australia (Figure 2). For instance, resistance of the Ridgelands and Biarra types has been shown to be due to the development of cholinesterases which are less sensitive to inhibition by OP-chemicals, resistance of the Mackay type to detoxication, whereas in resistance of the Gracemere and Mt. Alford types, detoxication has been added to insensitive cholinesterase (Wharton and Roulston, 1970; CSIRO 1972). The broad spectrum of resistance exhibited by these strains of ticks may be illustrated from the data of Roulston (1969) who recorded the resistance factors of Biarra strain ticks to 50 OP and carbamate chemicals which were existing or potential acaricides. The frequency distribution of resistance factors was as follows:

Factor	<2	2-5	5-10	10-50	>50
No. of chemicals	4	8	7	23	8

Thus the resistance factors to over three fourths of the chemicals were more than five, even though the ticks would have been subjected

Table 1

CHRONOLOGICAL RECORD OF ORGANOPHOSPHORUS (OP) RESISTANCE IN *BOOPHILUS MICROPLUS* IN AUSTRALIA IN RELATION TO INTRODUCTION AND USE OF OP AND FORMAMIDINE ACARICIDES (DDT BANNED IN 1962)

1956	First organophosphorus chemical (diazinon) introduced, followed by dioxathion (1958), coumaphos (1-59), ethion (1962) and the carbamate carbaryl (1963).
1963	Ridgelands (central Queensland) resistance to dioxathion recognised; cross-resistance to diazinon, carbophenothion and carbaryl.
1966	Biarra (south Queensland) resistance to coumaphos, ethion recognised; cross-resistance to almost all OP and carbamate chemicals, low resistance to prolate, bromophos ethyl and Dursban.
	Efficiency of chlordineform with different mode of action demonstrated.
1967	Mackay (north Queensland) resistance to same range of acaricides as Biarra recognised; toxicological (including higher resistance to carbamates and to prolate) and biochemical differences from Biarra resistance demonstrated.
	Dursban, bromophos ethyl, prolate introduced for field control.
1970	Mt. Alford (south Queensland) and Gracemere (central Queensland) resistance recognised; apparent extensions of, respectively, Biarra and Ridgelands resistance exhibiting enhanced resistance particularly to Dursban in both, and to bromophos ethyl in Mt. Alford.
1971	Additional types of resistance suspected, e.g. resistance to coumaphos but not to ethion.
	Chlordineform introduced for field control in mixtures with OP-chemicals.

to selection pressure by only one or two of the chemicals. Shaw, Cook and Carson (1968) in their report of the results of screening over 800 compounds presented similar data.

The area of greatest threat undoubtedly is southern Queensland where, since the recognition of Biarra resistance in 1966, there has been no acaricide available to control all ticks with the same efficiency and cost as a range of OP-chemicals could prior to its development. On many properties, control has come to depend either on the unstable OP-chemical prolate, the only OP-chemical to which Mt. Alford resistance is low, or on the formamidine, chlordineform. This chemical which is not a cholinesterase inhibitor was initially developed for use in mixtures with OP-acaricides (Roulston et al., 1971) but can now be used on its own and is probably the acaricide for problem situations. However, it is not an easy acaricide to use as the base is unstable, the chemical is converted to the acid salt, for use in a dipping vat, and the vat is buffered with superphosphate to a pH of about 5 (Roulston et al., 1971). The acaricidal potential of the formamidines was demonstrated in 1966 (Roulston and Wharton, 1967). It is hoped that other chemicals with a similar mode of action will be developed as acaricides. One such chemical, the diamidine BTS27419, demonstrates outstanding acaricidal activity (Palmer et al., 1973, Roulston, 1973).

Another chemical, clenpirin (Bayer 6896), an imino-pyrolidine, which is also not a cholinesterase inhibitor has been developed recently as an acaricide for the control of *Boophilus*; it is effective against OP-resistant strains but is inactive against three-host ticks and against insects. It also needs to be applied at the high concentrations of from 0.2 to 0.3 per cent. Consequently, the cost of treatment is high, three to four times as much as OP-acaricides used to control susceptible ticks.

There are also a few OP and carbamate chemicals which retain most of their activity against OP-resistant ticks. Promecarb (carbamult), which was shown by Roulston et al. (1971) to be fairly effective against Biarra, and subsequently against ticks of the Mt. Alford strain (W.J. Roulston, unpublished data) is one which may be developed as a commercial acaricide.

Fortunately resistance is not universal. High-level resistance of the Biarra or Mt. Alford types is very common in southern Queensland, but rare in coastal areas of central or northern Queensland, whereas in the western interior of Queensland, the Northern Territory and Western Australia, there is no evidence of resistance (Figure 2). Thus, there are many cattle owners who can still use dioxathion, coumaphos and ethion (and these acaricides still dominate the acaricide market). On the other hand, the extension in 1970-71 of Biarra resistance to northern New South Wales, where tick control is under direct government supervision

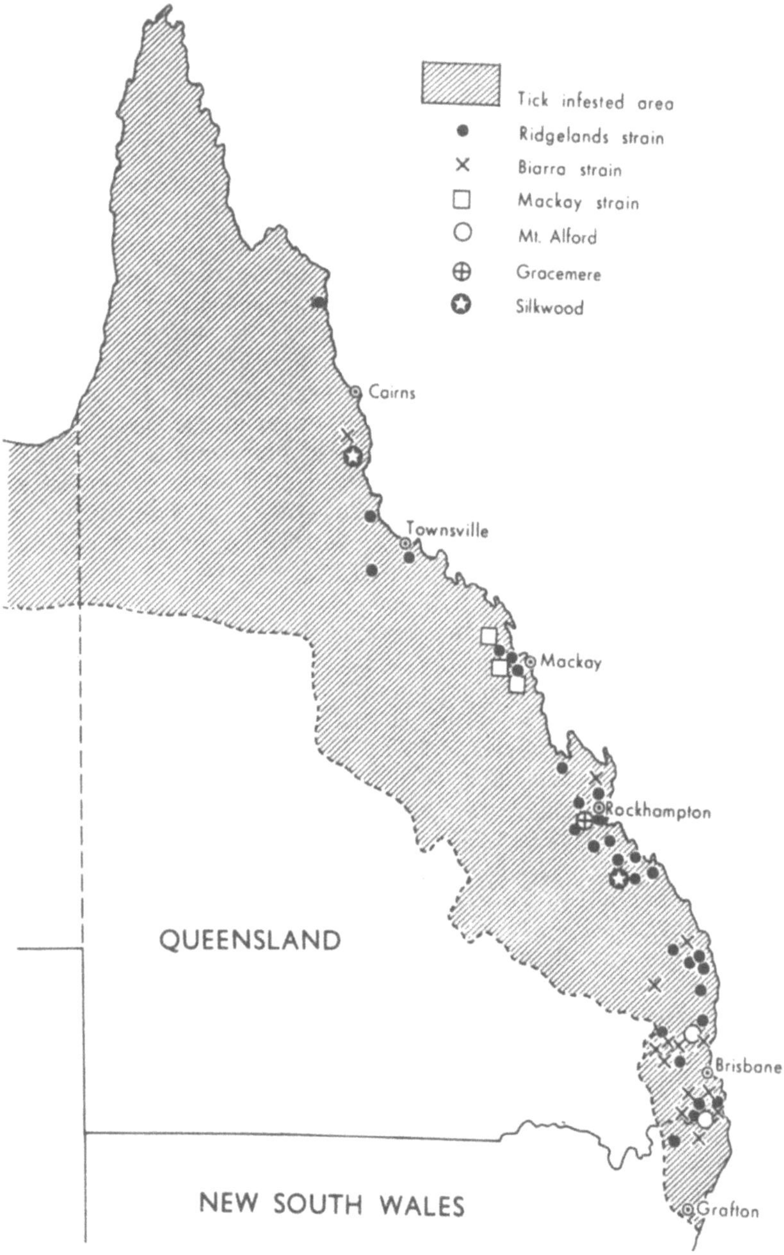
Tick infested area
Ridgelands strain
Biarra strain
Mackay strain
Mt. Alford
Gracemere
Silkwood
Cairns
Townsville
Mackay
Rockhampton
QUEENSLAND
Brisbane
NEW SOUTH WALES
Grafton

Figure 2.

and carried out in government-controlled and operated dipping vats, poses perhaps one of the greatest challenges to control authorities since the tick entered the state some 70 years ago. When arsenic-resistance was discovered in 1952, DDT was an obvious substitute acaricide (Legg and Shanahan, 1954). OP-acaricides replaced DDT in 1962 and have been used very successfully. Now there are no obvious alternatives. Biarra resistance can be controlled only by Dursban or bromophos ethyl (at increased cost), but Mt. Alford resistance is across the border in southern Queensland and cannot be controlled by either of these acaricides.* The problem in northern New South Wales is compounded by the fact that tick control has been maintained at such a high level that tick fever is not normally transmitted. Therefore, there is a population of .75 million non-immune cattle in a potentially enzootic area (McCulloch 1969).

Resistance to OP-acaricides in South and Central America

There is little published information on OP-resistance from areas occupied by *B. microplus* in American countries but, apart from one locality in Rio Grande de Sul, Brazil, resistance does not appear to have reached the same degree of complexity or pose as serious a threat as in Australia (R.H. Wharton, unpublished). Resistance of the Ridgelands type has been reported from Venezuela (Shaw, Cook and Carson, 1968) and almost certainly occurs in Argentina and Brazil. Other strains demonstrating increased resistance to coumaphos have been reported from Argentina (Flück and Rüfenacht, 1969; Torrado and Gutierrez, 1970). At the locality in Rio Grande de Sul, ticks have developed resistance in succession to dioxathion, ethion and Dursban (N.K. Amaral, personal communication) indicating that they now represent a strain with characteristics similar to those of Mt. Alford in Australia.

There is no record of OP-resistance in Central America. O.H. Graham (personal communication) believes that there is little or no resistance in Mexico.

ALTERNATIVE APPROACHES TO CONTROL

Pasture Spelling

This approach to control has been widely published in Australia since Wilkinson (1955) demonstrated that it is possible to control *B. microplus* by rotating cattle between pastures at three-to-five-month intervals. A similar approach was used extensively in the United States during the *B. annulatus* eradication campaign (Ellenberger and Chapin, 1919). There, however, pastures were left vacant for longer periods for eradication. Pasture spelling has

*See page 52.

had limited success for a variety of reasons: ignorance and apathy on the part of cattle owners; the cost of fencing and watering facilities, and difficulties in maximising pasture utilisation with resting and grazing periods, particularly under the extremely variable climatic conditions that prevail throughout northern Australia. A factor which is sometimes overlooked when considering the great attraction of this ecological approach to control, is the management of a cattle property with several breeding herds and the number of paddocks required. Thus, Waters (1972) describes the successful implementation of pasture spelling on a north Queensland coastal property with 2 500 cattle, including 1 000 breeding cows for which 15 paddocks are needed for breeding herds, weaner heifers and steers, and for bullocks. A considerable capital outlay was required to develop the facilities, but once established, the owner not only achieved excellent tick control, but also improved cattle management.

Resistant Cattle

The use of cattle which are resistant to tick infestation, expressed by their ability to prevent the maturing of large numbers of female *B. microplus* provides a second alternative approach to control. Resistance occurs in all breeds of cattle but is associated primarily with Zebu (*Bos indicus*) cattle (Siefert 1971; Wharton, Utech and Sutherst, 1973). The cattle tick, Zebu cattle and *Babesia* evolved together in Asia where they co-exist without obvious effects of the ticks or the *Babesia* on cattle. This association has been maintained in Brazil and other parts of Central America. However, despite the efforts of Kelley (1943) and others, little progress was made in introducing Zebu cattle into Australia where cattle in the tropics and sub-tropics were predominantly of British breeds (Shorthorn and Herefords)not adapted to the environment and requiring acaricidal treatments to avoid production losses. Demonstrations by Wilkinson (1962), Francis (1964), Wharton et al. (1969), Johnston and Haydock (1969), Gee, Bainbridge and Haslam (1971) have indicated the advantages that follow the use of resistant cattle; there is general acceptance that they provide a logical long-term solution to the cattle tick problem.

The fact that resistance is associated primarily with Zebu cattle was the main reason why this approach to tick control was not adopted earlier. However, at present, there is considerable prejudice as yet against these cattle and their crosses with European cattle, not only in Australia but also in southern Brazil and northern Argentina (R.H. Wharton, unpublished observations). Many cattle owners and scientists also find it unacceptable that the type of cattle to be used in an environment should be dictated by a parasite. Nevertheless, a major change in cattle owners' attitudes occurred in the 1960's in northern Australia. Zebu and Zebu x British cattle now dominate north of the tropic of Capricorn. The change has come about not only because of their resistance to ticks, but also because they are better adapted to the environment and can utilize the low quality pastures more efficiently. Regrettably, the change in attitude has yet to occur among the majority of cattle owners in the sub-tropical areas which are facing the most severe problems of acaricide resistance.

Although resistant cattle occur among European breed cattle (Wilkinson (1962) has demonstrated the improved control which can be achieved through selection for resistance within a herd), very little progress has been made towards breeding for tick resistance or inducing resistance in European breed cattle. Inheritable resistance (Wharton, Utech and Turner, 1970) and selection and breeding for tick resistance in a dairy Shorthorn herd is being attempted (CSIRO 1972). The results are encouraging with respect to the challenge to produce a tick-resistant herd, but the improvement is very slow compared with that achieved by introducing Zebu (*Bos indicus*) genes (Wharton, Utech and Sutherst, 1973, CSIRO 1972).

Unfortunately, the earlier encouraging results of attempts to induce resistance in European breed cattle by Riek (1962) have not been substantiated. The subsequent demonstration by Roberts (1968) that a major component of resistance is acquired and that each animal acquires its individual level of resistance, whether it be high or low means that naturally acquired levels of resistance will need to be enhanced if a programme aimed at inducing resistance in European breed cattle is to succeed.

PROSPECTS FOR THE FUTURE

The immediate prospects for the control of *B. microplus* are bleak in areas that are stocked by European cattle and are infested with the Biarra and Mt. Alford types of OP-resistance. As of the present, the costs of the acaricides applied are at least twice and, in some cases, four times as high as they were prior to OP-resistance. Alternative acaricides may be found, but they will undoubtedly be more expensive as the world market for acaricides is approximately US $20 million. Any new acaricide developed specifically for the control of OP-resistant ticks would serve only a fraction of the market. There is also the strong possibility that new acaricides cannot be found, in which case, the cattle owner would have to apply acaricides weekly instead of after intervals of three to four weeks. Progressive and intelligent cattle owners will apply their acaricides more efficiently and strategically in the spring in order to prevent a build-up of the summer and autumn populations (Norris 1957). Others will find it exceedingly difficult to maintain control and will demand alternative approaches to control.

Without wishing to be unduly pessimistic, the only approaches which merit serious consideration at this stage are pasture spelling, resistant cattle and chemicals.

Biological Control

In principle, biological control by parasites would be difficult to apply to the complex cattle, tick and *Babesia* association. Success is unlikely as biological control agents which are present in Asia, fail to protect introduced European cattle. Control would require the adaptation of known hymenopterous parasites of three-host ticks to the one-host *Boophilus* tick, or, if pathogenic bacteria or viruses, would require application to cattle in the same way as an acaricide and be subject to the same risks of resistance.

Control by predators, such as the ox-pecker tick birds of Africa, would be totally unacceptable on quarantine grounds, of very doubtful value and inadvisable because of potential competition with and possible displacement of the indigenous fauna.

Genetic Manipulations

Genetic manipulation, including the use of sterilized males or multiple translocations would appear to have no application for *Boophilus* control because of the dispersion that occurs almost exclusively on cattle which may carry thousands of ticks in all stages. Thus to achieve predominance or even parity of laboratory-reared ticks applied to cattle in the field would require at least equivalent numbers of cattle in the laboratory. Attempts to maintain *B. microplus in vitro* have failed and the prospects are poor.

Pasture Spelling

Pasture spelling will be applied more widely by the intelligent and progressive cattle owner, particularly as pastures improve and fodder conservation becomes a more practicable economic proposition.

Resistant Cattle

Resistant cattle are an attractive and logical approach and have already reduced the tick to an insignificant problem, as in Brazil, on many cattle properties in northern Australia. Their use offers an important and fascinating challenge in the field of population dynamics: different degrees of resistance (Figure 3) are

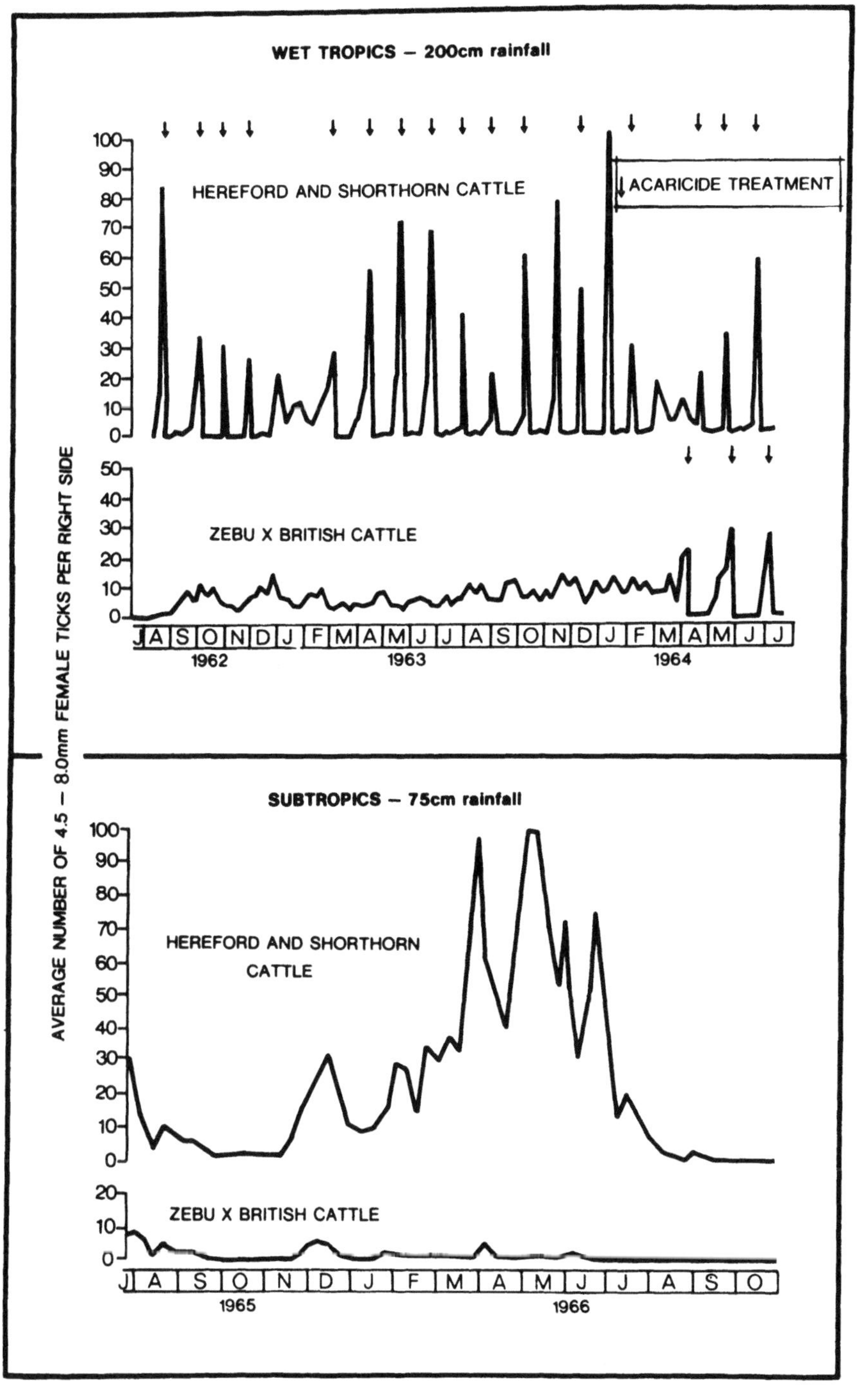
WET TROPICS – 200cm rainfall
HEREFORD AND SHORTHORN CATTLE
ACARICIDE TREATMENT
ZEBU X BRITISH CATTLE
AVERAGE NUMBER OF 4.5 – 8.0mm FEMALE TICKS PER RIGHT SIDE
J A S O N D J F M A M J J A S O N D J F M A M J J
1962
1963
1964
SUBTROPICS – 75cm rainfall
HEREFORD AND SHORTHORN CATTLE
ZEBU X BRITISH CATTLE
J A S O N D J F M A M J J A S O
1965
1966

Figure 3.

required to produce satisfactory tick control in different habitats (Wharton, Utech and Sutherst, in press). Reluctance to change and the real merit of European *Bos taurus* breeds of cattle, on their own but selected for the environment, or for cross-breeding programmes with Zebu cattle for heterosis, offers a further challenge to find economical methods of controlling ticks on these cattle, perhaps by breeding for resistance, perhaps by induction. Dairy cattle pose a special problem owing to the limited experience in the production of satisfactory *Bos taurus* × *Bos indicus* dairy cattle. Some progress has been made by Hewetson (1972); however, there is much less acceptance of the need by the dairy industry than by the beef cattle industry.

Chemical Control

While our major problems are associated with chemicals, it would be absurd to believe that control under all circumstances can be obtained without chemicals. Should eradication, within a single property, in an area of or throughout a country be attempted, chemicals must form the major basis of attack. There is a real threat of no alternative conventional chemicals because of acaricide-resistance. Clearly, it is time more thought and research effort were given to looking at the special physiological and biological characteristics of ticks which make them more vulnerable to attack by chemicals. The most obvious and perhaps the most promising would be the intimate and relatively long association of the feeding tick with its host. Thus a chemical which interfered with the "cement" used by the tick for attachment to its host, either during or after its production, or a chemical which irritated the tick, could lead to tick detachment. This would be equivalent to death in another arthropod as the chances of a parasitic tick which was interrupted in its feeding, regaining a suitable host would be very slight. Another approach for which some progress is being made is the elucidation of the changes in the cholinesterase which have occurred in OP-resistant ticks and the designing of cholinesterase-inhibitor chemicals which can cope with the altered cholinesterase (Anon., 1972).

Perhaps the most important aspect of chemical control is to learn how to conserve those which have been developed at such cost and with such skill and ingenuity. Personally, I have no doubt that *B. microplus* will respond to whatever chemical challenge it is given, if the challenge is sufficiently great and sufficiently constant. The concept of using chemicals to "live with the tick and tick fever" favours the development of resistance. We must, therefore, either eradicate the tick where this is economical and feasible or reduce our dependence on acaricides by the use of management practices and by resistant cattle.

REFERENCES

Anderson, J.L. (1963) Cattle tick (*Boophilus microplus*) - its occurrence and attempted eradication in the Territory of Papua and New Guinea, *Papua and New Guinea Agric. J.*, 15, 91-104

Anon. (1962) Report of Cooperative Cattle Fever Tick Eradication Activities USDA Fiscal Year 1962

Anon. (1972) Cooperative State-Federal Tick Eradication Activities USDA Fiscal Year 1971

Bolte, J.R., Hair, J.A. and Fletcher, J. (1970) White-tailed deer mortality following tissue destruction induced by lone-star ticks, *J. Wildlife Management*, 34, 546-552

Brocklesby, D.W. (1966) Theileridae of the buffalo in East Africa, *Proc. First Int. Cong. Parasit.*, 274-5

CSIRO (1972) Division of Entomology Annual Report for 1970-71, 103-104

Ellenberger, W.P. and Chapin, R.M. (1919) Cattle-fever ticks and methods of eradication, *U.S. Dept. Agr., Farmers' Bull.*, 1057

Flück, V. and Rüffenacht, K. (1969) Effectiveness of newer phosphorus compounds against resistant ticks of *Boophilus microplus*, Soc. Chem. Ind. Monograph No. 33, 183-193

Francis, J. (1964) Resistance of Droughtmaster cattle to tick infestation and babesiosis, *Aust. vet. J.*, 40, 247-253

Gee, R.W., Bainbridge, M.H. and Haslam, J.Y. (1971) The effect of cattle tick (*Boophilus microplus*) on beef production in the Northern Territory, *Aust. vet. J.*, 47, 257-263

Hair, J.A. and Howell, D.E. (1970) Lone-star ticks: their biology and control in Ozark recreational areas, *Oklahoma Agric. Expt. Sta. Bull.*, 106, 239

Hewetson, R.W. (1969) The inheritance of resistance by cattle to cattle tick, *Aust. vet. J.*, 48, 299-303

Hoogstraal, H. (1956) African Ixodoidea I. Ticks of the Sudan, *U.S. Naval Med. Res. Unit No. 3, Cairo*, 1101

Johnston, L.A.Y. and Haydock, K.P. (1969) The effect of cattle tick (*Boophilus microplus*) on production of Brahman-cross and British-breed cattle in northern Australia, *Aust. vet. J.*, 45, 175-179

Kelley, R.B. (1943) Zebu cross cattle in Northern Australia, *Bull. Coun. Sci. Ind. Res. Melb.*, No. 172

Legg, J. (1959) Report to the Government of the Federation of Malaya on animal health problems with special reference to protozoan blood parasites, FAO Report No. 1090, Rome 26 pp. (mimeographed)

Legg, J. and Shanahan, G.J. (1954) The appearance of an arsenic-resistant cattle (*Boophilus microplus*) in a small area of New South Wales. Its reaction to some of the newer insecticides, *Aust. vet. J.*, 30, 95-99

Mahoney, D.F. (1972) Epizootiological factors in the control of bovine babesiosis, *Aust. vet. J.*, 48, 292-298

Mackerras, I.M., Waterhouse, D.F., Maiden, A.C.B. and Edgar, G. (1961) The cattle tick problem in N.S.W., *Dept. Agr. N.S. Wales, Sci. Bull.*, 78, 100

McCulloch, R.N. (1969) Controlling the cattle tick in New South Wales, *Agric. Gazette of N.S.W.*, 80, 356-368

Norris, K.R. (1957) Strategic dipping for the control of the cattle tick (*Boophilus microplus*) Canestrini in south Queensland, *Aust. J. agric. Res.*, 8, 768-787

Palmer, B.H., McCarthy, J.F., Kozlick, A. and Harrison, I.R. (1973) A new chemical group of cattle acaricides, *Proc. 3rd Int. Congr. Acarol.*, *1971*, 687-692

Riek, R.F. (1962) Studies on the reactions of animals to infestation with ticks. VI. Resistance of cattle to infestation with the tick *Boophilus microplus* (Canestrini), *Aust. J. agric. Res.*, 13, 532-50

Riek, R.F. (1965) The cattle tick and tick fever, *Aust. vet. J.*, 41, 211-215

Roberts, F.H.S. (1965) The taxonomic status of the species of the genera *Rhipicephalus* Koch and *Boophilus* Curtice (Acarina: Ixodidae) occurring in Australia, *Aust. J. Zool.*, 13, 491-523

Roberts, J.A. (1968) Acquisition by the host of resistance to the cattle tick, *Boophilus microplus* (Canestrini), *J. Parasit.*, 54, 657-662

Roulston, W.J. (1969) Acaricide resistance in the cattle tick, *Boophilus microplus* in Australia, *Proc. 2nd Int. Congress Acarol.*, *1967*, 515-521

Roulston, W.J. (1973) Prospects for chemical control of *Boophilus microplus* in Australia, *Proc. 3rd Int. Congress Acarol.*, *1971*, 693-696

Roulston, W.J. and Wharton, R.H. (1967) Acaricide tests on the Biarra strain of organophosphorus-resistant cattle ticks *Boophilus microplus* from southern Queensland, *Aust. vet. J.*, 43, 129-34

Roulston, W.J., Wharton, R.H., Schnitzerling, H.J., Sutherst, R.W. and Sullivan, N.D. (1971) Mixtures of chlorphenamidine with other acaricides for the control of organophosphorus-resistant strains of cattle tick *Boophilus microplus*, *Aust. vet. J.*, 47, 521-528

Seifert, G.W. (1971) Variations between and within breeds of cattle in resistance to field infestations of the cattle tick (*Boophilus microplus*), *Aust. J. agric. Res.*, 20, 783-797

Shaw, R.D., Cook, M. and Carson, R.E. (1968) Developments in the resistance status of the southern cattle tick to organophosphorus and carbamate insecticides, *J. Econ. Ent.*, 61, 1590-94

Theiler, G. (1958) First meeting of the joint FAO/OIE expert panel on tick-borne diseases of livestock. Predators - a survey of the literature, 94-100

Torrado, J.M.L. and Gutierrez, R.O. (1970) Fosforo-resistencia de une cepa argentina de garrapata *Boophilus microplus*, *Rev. medicina Veterinaria*, 51, 3-12

Ulloa, G. and de Alba, J. (1957) Resistencia a los parasitos externos en elgunas razas de bovinos, *Turrialba*, 7, 8-12

Walton, G.A. (1962) The *Ornithodoros moubata* superspecies problem in relation to human relapsing fever epidemiology, *Symposium Zool. Soc. Lond.*, 6, 83

Waters, K. (1972) Pasture spelling for the tick, *Queensland Agric. J.*, 98, 170-175

Wharton, R.H., Harley, K.L.S., Wilkinson, P.R., Utech, K.B.W. and Kelley, B.M. (1969) A comparison of cattle tick control by pasture spelling, planned dipping and tick-resistant cattle, *Aust. J. agric. Res.*, 20, 783-97

Wharton, R.H., Utech, K.B.W. and Turner, H.G. (1970) Resistance to the cattle tick, *Boophilus microplus* in a herd of Australian Illawarra Shorthorn cattle - its assessment and heritability, *Aust. J. agric. Res.*, 21, 163-181

Wharton, R.H. and Roulston, W.J. (1970) Resistance of ticks to chemicals, *A. Rev. Ent.*, 15, 381-404

Wharton, R.H. and Utech, K.B.W. (1970) The relation between engorgement and dropping of *Boophilus microplus* (Canestrini) (Ixodidae) to the assessment of tick numbers on cattle, *J. Aust. ent. Soc.*, 9, 171-182

Wharton, R.H., Utech, K.B.W. and Sutherst, R.W. (1973) Tick resistant cattle for the control of *Boophilus microplus*, *Proc. 3rd Int. Congress Acarol.*, *1971*, 697-700

Wilkinson, P.R. (1955) The spelling of pasture in cattle tick control, *Aust. J. agric. Res.*, 8, 414-423

Wilkinson, P.R. (1962) Selection of cattle for tick resistance, and the effect of herds of different susceptibility on *Boophilus* populations, *Aust. J. agric. Res.*, 13, 974-83

Wilkinson, P.R. (1970) Factors affecting the distribution and abundance of the cattle tick in Australia: observations and hypotheses, *Acarologia*, 12, 492-508

* Mixtures of chlordimeform with organophosphorus acaricides or chlordimeform alone both buffered with superphosphate, butyryl promecarb or dipofene have proved to be satisfactory alternative acaricides to control Mt Alford organophosphorus-resistant *B. microplus* but are less efficient and more expensive than organophosphorus acaricides were against susceptible ticks.

DISCUSSION OF THE PAPER ON TICKS WITH SPECIAL EMPHASIS ON *BOOPHILUS MICROPLUS*

R.O. Drummond

U.S. Livestock Insects Laboratory

Agr. Res. Serv., USDA, Kerrville, Texas 78028

I do appreciate Dr Wharton's furnishing me a copy of his paper several weeks before this meeting so that I could formulate these comments. He and his entire research team are to be congratulated on the excellence of the research that has emanated in recent years from the CSIRO laboratory at Indooroopilly, Queensland, Australia.

The paper presented by Dr Wharton contains an excellent review of the scope of resistance of ticks, especially *Boophilus microplus* (Canestrini) in Queensland, to acaricides and also summarizes much of the research devoted to finding solutions for this problem. As I see it, the problem in Queensland is that stockmen have accepted for many years certain ranching practices which have possibly induced and certainly contributed to the problem. Among these practices are (1) incomplete control of *B. microplus*, (2) raising *Bos taurus* cattle rather than *Bos indicus* cattle, and (3) grazing cattle in large, relatively wild, pastures and thus not inspecting cattle frequently, if at all. The stockmen's preference to "live with the tick and the disease" was and remains a dominating influence on ranching practices. Rather than eradicate *B. microplus* and thus create large populations of cattle that would be highly susceptible to babesiosis, the cattlemen preferred to keep low populations of the tick and thus insure resistance to babesiosis in all cattle. This procedure would be unnecessary if an effective vaccine were available.

The paper points out several methods of combating the acute problem of resistance of *B. microplus* to acaricides. Briefly, the most promising of these methods are (1) judicious and proper use of presently available and new highly effective acaricides, (2) breeding of tick-resistant cattle, and (3) rotation of cattle to prevent pastures from becoming highly infested with ticks. In the past, and presently, the main method of control of *B. microplus* has been and is the use of acaricides. The paper describes the dynamics of the increasing resistance in *B. microplus* to acaricides and draws a bleak picture of the development and use of new acaricides in the future. The paper suggests that the problem of resistance may force stockmen in Queensland and other areas to rely more on breeding and rotation for the control of *B. microplus*.

In reviewing the paper by Dr Wharton, I was struck by the following words: ignorance, apathy, emotional, prejudice, attitudes and reluctance. These are not words usually found in scientific papers, but I believe their use is quite valid in describing the situation as Dr Wharton finds it in Queensland. Research workers in agricultural sciences, especially those employed by government agencies, have a responsibility to the taxpayers who are the source of their funds. Money from the taxpayers is invested in research programmes which will eventually provide answers to their problems. It seems to me that research on *B. microplus* in Queensland has developed to the point where several solutions are available to solve the problem; however, because of attitudes that are described by words like ignorance, it has been difficult to implement these solutions. In my opinion, it is time for the results of the research in Queensland to be made clearly known to the rancher. If the researchers can demonstrate that the problems caused by resistance of *B. microplus* to acaricides can be solved by correct use (strategic dipping) of presently available and new acaricides and by alternative methods of control such as resistant cattle and pasture spelling, then they must make the cattlemen clearly aware of their findings. To convince the cattlemen, I believe the solution must have four basic characteristics: it must be Logical, it must be Enduring, it must be Attractive, and it must be Decisive. If these words are taken together, they illustrate the point I wish to make, that is, the researcher must now lead the rancher to solve the problem. Indeed, this leading may have to take the form of legislation that makes it obligatory for all stockmen to follow certain rules and regulations concerning cattle management and treatment. After all, we in publicly-funded research are hired by the people to solve their problems, and in solving their problems, we are only doing our job.

Finally, I still believe that *B. microplus*, since it is a one-host species and is limited to bovines, can be eradicated through a scheme of compulsory dipping of cattle in an acaricide that will kill 99 per cent of the ticks on the animal. Such a Government-conducted programme should succeed if the ranchers were thoroughly educated and motivated so the eradication programme will be supported by all of the participants. The costs of the programme might be high, but they would be relatively minor compared with the large losses that can be anticipated when previously available acaricides are no longer effective against *B. microplus*.

SIMULIUM DAMNOSUM

R. Le Berre

Office de la Recherche Scientifique

et Technique Outre Mer, Paris

Simulium damnosum is the main African vector of onchocerciasis. The name *S. damnosum*, which is used throughout the present paper for the sake of simplicity, covers in fact a relatively large number of forms whose taxonomic status is yet uncertain. Most observations reported here deal with the *S. damnosum* complex forms found in West Africa and, especially, with those presently called "Sirba", "Nile", "Soubré" and "Bandama".

The females of *S. damnosum* lay their eggs in batches of 250 at water level or up to a few millimetres below the surface, on partially submerged supports in fast-flowing rivers and their large tributaries. Various types of supports are used, including tree branches, plants and rocks. The eggs hatch between 36 and 48 hours after laying. The young larvae immediately move down to the submerged part of the support; they may also drift downstream and colonize completely immersed supports. The larvae, which generally inhabit the upper 50 cm of the river's surface water, feed by filtering organic and other particles carried by the current. *S. damnosum* is a fairly demanding species and will only inhabit those parts of the watercourse with a relatively high content of suspended nutritive particles. Under the prevailing conditions of most of the lowlands of tropical Africa, the larval life-cycle lasts for some eight to ten days. At the end of this period, the larvae moult into pupae after surrounding themselves with cocoons attached to a submerged support. The pupae neither move nor feed; this stage lasts about three days. The adults emerge from the pupae in an air bubble and take flight as soon as they reach the surface of the water.

The females of *S. damnosum* mate once in their life within a few hours of their emergence. The new-born females feed on plant juices before taking a blood meal, usually the day after their emergence. Each blood meal is followed by the maturation of the eggs to be laid three to five days later and each laying is followed within 24 hours by a new blood meal; these cycles continue in the same way until the death of the female. Female *S. damnosum* bite only by day.

Little is known at present about the adult resting places other than that just after emergence or a short while before laying, when the unfed and gravid females are sometimes found in large numbers on vegetation near the larval sites. The females may travel great distances when climatic conditions and vegetation are favourable. Their active dispersion may occur very rapidly over a distance of up to 40 km around the larval breeding sites throughout the year in rain forests, and during the rainy season in the savanna. On the other hand, in the dry season in the savanna, the adults of *S. damnosum* remain in the immediate vicinity of the larval breeding sites, going further only where gallery forests afford cover. *S. damnosum* females may also travel distances of up to 150 km or more with favourable air currents, especially those occurring with the shift of the inter-tropical front from the coast towards the interior at the beginning of each rainy season. These long flights ensure the annual repopulation of *S. damnosum's* temporary larval breeding sites in the savanna, when the watercourses flow again.

S. damnosum is attacked by a number of predators, parasites and pathogens. The use of mermithid parasites for controlling the vector is under study. However, it will probably be many years before this method becomes operational. Furthermore, these parasites affect only a certain proportion of the *Simulium* population and the anticipated degree of control is presently unpredictable, although a fair reduction in numbers can be expected.

Because of their ecological requirements (food, current, immersed supports), *S. damnosum* larvae occupy restricted stretches of certain rivers and are plentiful in a limited number of sites. The mechanical destruction of these major larval sites has been considered. The simplest solution would be the construction of dams downstream to transform fast flowing waters into still reservoirs. It must also be remembered that each dam spillway may provide an excellent *S. damnosum* breeding site unless properly designed. *S. damnosum* distribution has already been considerably expanded by a series of small dams set up to meet farming and ranching needs; these have generated far more breeding sites than they have destroyed. On the other hand, large dams destroy many more *S. damnosum* breeding sites than they produce. This method, while effective, is costly and can only be considered if such a dam is economically justifiable.

The use of insecticides is at present, therefore, the only practical means of controlling *S. damnosum*. Control of adults might be possible, but would be difficult owing to their rapid dispersal over vast areas. Larval control is relatively easy and more effective as the breeding sites are restricted to certain areas. The application of insecticides,upstream from the breeding sites to be destroyed,enables the toxic particles to be carried by the current to the larvae which absorb them in filtering their food, thus ensuring a concentration of the toxicant within their body. Owing to these characteristics, *Simulium* larvae are much more susceptible to insecticides than the great majority of other aquatic organisms, and their selective destruction is possible. The development of larvicide formulations that will remain in the upper layer of water also means that the toxicant can be concentrated in that part of the river where *S. damnosum* larvae are to be found, increasing further the selective aspect of the treatment.

Up to the present, the campaigns undertaken against *S. damnosum* in different parts of Africa have been based on the use of DDT, weekly applications of which gave the best results. To avoid all risk of long-term contamination of the environment, DDT is now being replaced by bio-degradable insecticides that disappear spontaneously from the treated areas a few weeks after application. The new compounds, which are not toxic for man in routine use, have also been selected for their lack of acute toxicity for non-target freshwater organisms, particularly fish and their natural food-chain, at concentrations which provide effective control of *S. damnosum* larvae (about 0.05 to 0.10 ppm of active ingredient, during 10 minutes, every week, which corresponds to an average concentration in the water of the treated rivers of some 0.05 to 0.1 mg per cubic metre of water). Thus the elimination of *S. damnosum* larvae need not have far-reaching effects on the biological equilibrium of the inland waters and will not interfere with the productivity of fisheries.

From a logistic viewpoint, the *S. damnosum* control operations undertaken in the Abuja and Kainji regions of Nigeria demonstrated that small-scale projects are hampered by the constant re-infestation of the treated area by female flies coming from the untreated neighbouring areas. The FED/OCCGE control project, covering about 60,000 sq. km. in the Ivory Coast, Mali and Upper Volta, showed conclusively that a large-scale scheme offers much better prospects of success than a small one. Furthermore, the cost of operations was less in proportion to the size of the project area. However, the implementation of a very large-scale *S. damnosum* control programme over hundreds of thousands of square kilometres raises logistic problems difficult to overcome, whether in the field of geographical reconnaissance, insecticide application, or the accurate evaluation of results.

The selection of the reaches to be treated, initially based on surveys made on foot, by boat, or motor vehicle, is now currently undertaken by light planes and helicopters, thus considerably accelerating reconnaissance operations, without any decrease in accuracy when helicopters are used. Operational research is in progress, under the guidance of the World Health Organization, to investigate the relative efficacy of fixed-wing aircraft and helicopters for the aerial application of larvicides, as well as to screen and evaluate new insecticides and to develop better formulations and specialized equipment. The combination of aerial treatment with an extensive ground network of capture stations for the permanent evaluation of the results would permit control operations for *S. damnosum* to cover very large areas. An ambitious programme of this type is under preparation in the Volta River basin area in West Africa, under the aegis of several United Nations agencies, involving seven countries and an area of approximately 700,000 square kilometres of Sudano-Guinean savanna. The aims of this programme are to prevent onchocerciasis transmission and to improve the economic development of the fertile valleys which have been deserted due to the severity of the clinical manifestations of this disease which, in the savanna belt, often culminates in blindness.

REFERENCES

Alekseev, A.N. (1963) Toxicity of some organophosphorous insecticides to larvae of bloodsucking Simuliids, *Med. Parazit. (Mosk.)*, 32, 546-548

Anon. (IDRC) (1972) Preventing onchocerciasis through blackfly control. A proposal for Afro-Canadian research into the feasibility of using mermithid parasites as biological agents in the control of disease-transmitting blackflies, *Publ. IDRC-006e*, Ottawa, 11 pp.

Barnley, G.R. (1953) Control of *Simulium damnosum* Theobald on the Victoria Nile, Uganda, WHO/ONCHO/18, 34 pp. (Unpublished document)

Browne, S.G. (1962) Rapport sur une campagne pilote antisimulidienne au Congo, *Ann. Soc. belge Méd. trop.*, 42, 847-853

Carlsson, G. (1971) Les parasites en tant que moyens de contrôle biologique pour les diptères d'importance médicale. Résumé de discussion, *Comptes rendus 1er Multicolloque européen Parasit.*, 412-415

Davies, J.B. (1963) An assessment of the insecticidal control of *Simulium damnosum* Theobald in Abuja Emirate, Northern Nigeria, from 1955 to 1960. I. The effect on the prevalence of onchocerciasis in the human population, *Ann. trop. Med. Parasit.*, 57, 161-181

Davies, J.B. (1965) An assessment of the insecticidal control of *Simulium damnosum* Theobald in Abuja Emirate, Northern Nigeria, from 1955 to 1960. II. The effect on the incidence of *Onchocerca* larvae in the vector, *Ann. trop. Med. Parasit.*, 59, 43-45

Davies, J.B., Crosskey, R.W., Johnston, M.R.L. and Crosskey, M.E. (1962) The control of *Simulium damnosum* at Abuja, Northern Nigeria, *Bull. Wld Hlth Org.*, 27, 491-510

Dunbar, R.W. and Vajime, C.G. (1972) The *Simulium (Erwardsellum) damnosum* complex: A report on cytotaxonomic studies up to April 1972, WHO/ONCHO/72.100, 13 pp. (Unpublished document)

Fredeen, F.J.H. (1970) A constant-rate liquid dispenser for use in blackfly larviciding, *Mosquito News*, 30, 402-405

Garnham, P.C.C. and McMahnon, J.P. (1954) Final results of an experiment on the control of onchocerciasis by eradication of the vector, *Bull. ent. Res.*, 45, 175-176

Hill, G. (1959) Methods of water-flow measurement applicable to *Simulium* control, *Bull. Wld Hlth Org.*, 21, 201-205

Hitchen, C.S. and Goiny, H.H. (1966) Note on the control of *Simulium damnosum* in the region of the Kainji Dam Project, Northern Nigeria, WHO/ONCHO/66.43, 5 pp. (Unpublished document)

Jamnback, H., Duflo, T. and Marr, D. (1970) Aerial application of larvicides for control of *Simulium damnosum* in Ghana, a preliminary trial, *Bull. Wld Hlth Org.*, 42, 826-828

Jamnback, H. and Higgins, A.E.H. (1969) The feasibility of aircraft spraying for *Simulium* larval control in West Africa., VBC/69.5, 12 pp. (Unpublished document)

Jamnback, H. and West, A.S. (1970) Decreased susceptibility of blackfly larvae to p,p'-DDT in New York State and Eastern Canada, *J. econ. Ent.*, 63, 218-221

Jannssens, P.G. (1968) Contrôle entomologique et thérapeutique de l'onchocercose, *Rapp. final 8e Conf. techn. OCCGE*, 1, 228-231, (Unpublished document)

Kuzoe, F.A.S. and Hagan, K.B. (1967) The control of *Simulium damnosum* Theobald (Diptera, Simuliidae) in the region of the Volta Dam, *Ann. trop. Med. Parasit.*, 61, 338-348

Le Berre, R. (1966) Contribution à l'étude biologique et écologique de *Simulium damnosum* Theobald, 1903 (Diptera, Simuliidae), *Mém. ORSTOM*, 17, 204 pp.

Le Berre, R. (1967) Possibilité et modalités de réalisation d'une campagne régionale de lutte contre le vecteur de l'onchocercose, 316/Oncho, 9 pp. (Unpublished document)

Le Berre, R. (1968a) Rapport sommaire sur le traitement de la zone de Tiassalé par voie aérienne; 7/ONCHO, 4 pp. (Unpublished document)

Le Berre, R. (1968b) La lutte centre le vecteur de l'onchocercose humaine, *Simulium damnosum* Th. en Afrique occidentale francophone; 198/ONCHO, 5 pp. (Unpublished document)

Le Berre, R. (1970) Résultats des campagnes de lutte, *Rapp. final 10e Conf. techn. OCCGE*, 1, 199-201, (Unpublished document)

Le Berre, R. (1972) Fiche technique concernant les répercussions de différents projets de barrage sur le Haut-Bandama (Région de Korhogo-Ferkessédougou); 6/ONCHO, 5 pp. (Unpublished document)

Le Berre, R. and Subra, R. (1964) Essais de traitements larvicides contre *S. damnosum* Theo. dans la région de Tiassalé, (Unpublished OCCGE document)

Lea, A.O. and Dalmat, H.T. (1954) Screening studies of chemicals for larval control of blackflies in Guatemala, *J. econ. Ent.*, 47, 135-141

McMahon, J.P. (1967) A review of the control of *Simulium* vectors of onchocerciasis, *Bull. Wld Hlth Org.*, 37, 415-430

McMahon, J.P., Highton, R.B. and Goiny, H. (1968) The eradication of *Simulium neavei* from Kenya, *Bull. Wld Hlth Org.*, 19, 75-107

Muirhead-Thomson, R.C. (1957) Laboratory studies on the reactions of *Simulium* larvae to insecticides. I. A laboratory method for studying the effects of insecticides on *Simulium* larvae, *Am. J. trop. Med. Hyg.*, 6, 920-925

Muirhead-Thomson, R.C. and Merryweather, J. (1969) Effect of larvicides on *Simulium* eggs, *Nature (Lond.)*, 221, 858-859

Muirhead-Thomson, R.C. and Merryweather, J. (1970) Ovicides in *Simulium* control, *Bull. Wld Hlth Org.*, 42, 174-177

Noamesi, G.K. (1962) A preliminary study of the susceptibility of *Simulium* larvae to DDT, y-BHC and dieldrin, *Bull. Wld Hlth Org.*, 27, 620-622

Noamesi, G.K. (1968) The strategy and planning of onchocerciasis control in Ghana; Joint USAID/OCCGE/WHO Techn. Meet. on the Feasibility of Onchocerciasis Control, ONCHO/WP/68.9, 26 pp., (Unpublished document)

Noamesi, G.K. (1970) Preliminary report on a new repellent "Simno" against blackflies and other biting insects, *Ghana med. J.*, 9, (1), 39-43

Ovazza, M. and Valade, M. (1963) Recherches sur la prophylaxie de l'onchocercose humaine en Afrique de l'ouest de langue française. II. Essais de larvicides sur le terrain et au laboratoire, *Bull. IFAN*, 25, série A, 1215-1234

Perti, S.L. and Lopez, W.A. (1964) Bionomics and control of *Simulium* flies in Nefa area. Part II - Susceptibility of *S. indicum* larvae to insecticides, *Armed Forces med. J.*, 20, 180-188

Perti, S.L. and Lopez, W.A. (1964) Bionomics and control of *Simulium* flies in Nefa area. Part III - Field trials on the control of *S. indicum*, *Armed Forces med. J.*, 20, 189-193

Philippon, B. (1971) Etude entomologique du foyer d'onchocercose de la vallée du Ouassoulou (Cercle du Yanfolila - République du Mali). Résultats des enquêtes - Avant-projet de traitement antisimulidien, 141/ONCHO, 9 pp. (Unpublished document)

Quelennec, G. (1964) Projet d'organisation d'une zone de contrôle de *Simulium damnosum* Theobald 1903 dans le nord-ouest Dahomey. (Unpublished OCCGE document)

Quelennec, G. and Ovazza, M. (1968) La lutte contre les simulies *Cah. ORSTOM Ser. Ent. med.* 6, 197-205

Quelennec, G., Simonkovich, E. and Ovazza, M. (1968) Recherche d'un type de déversoir de barrage défavorable à l'implantation de *Simulium damnosum* (Dipt. Sim.), *Bull. Wld Hlth Org.*, 38, 943-956

Quelennec, G. and Vervent, G. (1970) Mesure de la sensibilité aux insecticides des larves de Simulies (Dipt. Sim.), *Cah. ORSTOM, sér. Ent. méd.*, 8, 21-44

Quillevere, D., Escaffre, H., Pendriez, B. et al. (1972) Lutte contre *Simulium damnosum* vecteur de l'onchocercose humaine en Afrique occidentale. III. Epandages par avion en saison des pluies. Méthodes d'épandage. Nouvelles formulations. 174/ONCHO, 12 pp. (Unpublished document)

Quillevere, D., Escaffre, H., Pendriez, B. et al. (1973) Lutte contre *Simulium damnosum* vecteur de l'onchocercose humaine en Afrique occidentale. IV. Expérimentation par hélicoptère de nouvelles formulations et simulation d'une campagne insecticide; 1/ONCHO, 29 pp. (Unpublished document)

Roberts, J.M.D., Neumann, E., Gockel, G.W. and Highton, R.B. (1967) Onchocerciasis in Kenya 9, 11 and 18 years after elimination of the vector, *Bull Wld Hlth Org.*, 37, 195-212

Shvankov, M.E. (1966) Economic comparison of different methods of bloodsucking blackflies (Simuliidae) control in the Krasnoyarsk Region, *Med. Parazit. (Mosk.)*, 4, 498-499

Sow, C. (1970) Campagnes de lutte FED/OCCGE contre l'onchocercose. Bilan financier, *Rapp. final 10e Conf. techn. OCCGE*, 1, 204-208, (Unpublished document)

Suzuki, T., Ito, Y. and Harada, S. (1963) A record of blackfly larvae resistant to DDT in Japan, *Jap. J. exp. Med.*, 33, 41-46

Taufflieb, R. (1955) Une campagne de lutte contre *Simulium damnosum* au Mayo Kebbi, *Bull. Soc. Path. exot.*, 48, 564-576

Taufflieb, R. (1956) Rapport sur la campagne antisimulidienne de 1956 au Mayo Kebbi, *Bull. Inst. Et. centrafr.*, (n.s.), 11, 53-59

Waddy, B.B. (1969) Prospects for the control of onchocerciasis in Africa with special reference to the Volta River basin, *Bull. Wld Hlth Org.*, 40, 843-858

Walsh, J.F. (1970a) Evidence of reduced susceptibility to DDT in controlling *Simulium damnosum* (Diptera: Simuliidae) on the River Niger, *Bull. Wld Hlth Org.*, 43, 316-318

Walsh, J.F. (1970b) Lutte antivectorielle contre *Simulium damnosum* le long du Niger et de ses affluents en liaison avec le projet de recherches concernant le lac Kainji pendant la période 1961-1969, PD/70.4, 48 pp. (Unpublished document)

Wanson, M. (1950) Contribution à l'étude de l'onchocercose africaine humaine (problèmes de prophylaxie à Léopoldville), *Ann. Soc. belge Méd. trop.*, 30, 667-863

Wanson, M., Courtois, L. and Lebied, B. (1949) L'éradication du *Simulium damnosum* Theobald à Léopoldville, *Ann. Soc. belge Méd. trop.*, 29, 373-403

Wilson, C. (1956) Combatting onchocerciasis with little expense, *W. Afr. med. J.*, (n.s.), 5, 162

WHO (1970) Insecticide resistance and vector control. Seventeenth report of the WHO Expert Committee on Insecticides, *Wld Hlth Org. techn. Rep. Ser.*, 443, 10 et 120-124

WHO (1971) IV. Blackflies in the Americas, WHO/VBC/71.283, 24 pp. (Unpublished document)

WHO (1973) Informal consultation on *Simulium damnosum* control in the Volta River basin area, West Africa, Geneva, 12-15 December 1972, Report, VBC/OCP/73.1, 15 pp. (Unpublished document)

DISCUSSION OF THE PAPER ON *SIMULIUM DAMNOSUM*

D.J. Lewis

Scientific Staff, Medical Research Council London

British Museum (Natural History)

In the first decade of this century, *Simulium damnosum* was discovered; in the third, it was recognized as a vector of human onchocerciasis. In the fourth decade, this disease, and in many areas river blindness, was accepted as an inevitable hazard, receiving much less attention than the killing diseases, while a few enthusiasts gathered knowledge on the vector. The fifth decade was one of promise, with the advent of DDT and some spectacular successes against *Simulium* larvae. Between 1951 and 1960, new areas were put under control including the Abuja region of Nigeria by R.N. Crosskey (1958). The seventh decade was one of increasing knowledge and of appraisal of results. *S. damnosum*, once regarded as a rather uniform species, was seen to be a complex of cytological categories and morphological variants. Physiological differences between some of these variants were revealed by B.O.L. Duke's discovery of their association with geographical forms of *Onchocerca volvulus*.

S. damnosum taken on human bait in West African forests were found to have a lower average parous rate than in the savanna where flies apparently tend to live longer and blindness is a serious problem. Interpretation of the relation between observed parous rates and transmission of *O. volvulus* proved complex in the light of the findings of R.H.L. Disney (1972) in West Africa of a degree of zoophily which may vary with age and can influence trap catches. Such questions and the existence of non-anthropophilic populations of the vector in East Africa showed how much remained to be learned about the basic biology of this species or complex. As J.B. Davies (1965) pointed out, action against *S. damnosum* was not always followed by adequate control of transmission probably owing to the great flight range of females which could penetrate deep into areas treated with larvicide. This serious problem could not be solved by combining simuliid control with medication, as the basic curative drugs were not suitable for mass treatment. The solution as indicated by the work of R. Le Berre seemed to be a costly enlargement of control areas despite lack of roads and other difficulties.

At the beginning of the eighth decade, we see a paradox. A simple method of vector control, the application of small quantities of larvicide to restricted breeding places, has long been known, and yet most of the African onchoceriasis area is untreated. The rural nature of the problem involves great logistic and financial difficulties.

The present situation just reviewed by Dr Le Berre, is one of modified optimism. A great fund of knowledge is available, and the World Health Organization has carried out extensive planning, and a massive international effort is envisaged. This will include the use of aircraft to cover large areas rapidly, not just as an emergency measure but for a continuing programme for the control of *Simulium*. Groundwork will still be necessary, however, and the interactable problem of logistics will always be present.

Many breeding places of *S. damnosum* are in nearly flat open country suitable for aerial spraying. The need for this is enhanced by the existence of dry-season breeding sites revealed by G.K. Noamesi (1971) in Ghana which are apparently too numerous for an easy attack on the vector at this period. Jamnback et al. (1970) have pointed out that aerial spraying would permit the use of oil solutions which are less harmful than emulsions to non-target organisms.

S. damnosum is attacked by several parasites and predators Marshall Laird and his colleagues are planning field trials of control by mermithid nematodes. In future decades, the problems will be to consolidate results on a permanent basis at a reasonable cost, to prevent re-infestation from uncontrolled districts and to extend control to the vast untouched areas which include wooded uplands where aerial spraying may be difficult and to districts with poor soil and rainfall which cannot afford control.

In short, the challenge of *S. damnosum* control is a conflict of assets and liabilities of knowledge and its potentials, and of formidable practical obstacles.

REFERENCES

Crosskey, R.W. (1958) First results in the control of *Simulium damnosum* Theobald (Diptera, Simuliidae) in Northern Nigeria, *Bull. ent. Res.*, 49, 715-735

Davies, J.B. (1965) An assessment of the insecticidal control of *Simulium damnosum* Theobald in Abuja Emirate, northern Nigeria, from 1955 to 1960, II. The effect on the incidence of *Onchocerca* larvae in the vector, *Ann. trop. Med. Parasit.*, 59, 43-46

Disney, R.H.L. (1972) Observations on chicken-biting blackflies in Cameroon with a discussion of parous rates in *Simulium damnosum*, *Ann. trop. Med. Parasit.*, 66, 149-158

Duke, B.O.L., Lewis, D.J. and Moore, P.J. (1966) *Onchocerca-Simulium* complexes. I. Transmission of forest and Sudan-savanna strains of *Onchocerca volvulus*, from Cameroun, by *Simulium damnosum* from various West African bioclimatic zones, *Ann. trop. Med. Parasit.*, 60, 318-336

Jamnback, H.A., Duflo, T. and Marr, D. (1970) Aerial application of larvicides for control of *Simulium damnosum* in Ghana: a preliminary trial, *Bull. Wld Hlth. Org.*, 42, 826-828

Le Berre, R. (Lecture in the Congress) See Chapter 6

Naomesi, G.K. (1971) Dry-season survival of *Simulium damnosum* Theobald 1903 (Diptera: Simuliidae) in northern Ghana, West Africa, *Ann. trop. Med. Parasit.*, 65, 555-564

STEGOMYIA

G.A.H. McClelland

Department of Entomology

University of California, Davis, California

For the purposes of this paper, I shall define control as any amount of reduction in the size of an insect population resulting from deliberate human interference. I shall largely restrict my discussion to four target species: *Aedes aegypti*, *Aedes simpsoni*, *Aedes albopictus* and *Aedes polynesiensis*, the first of which will receive a disproportionate share of attention.

Control of *A. aegypti* began in Cuba in 1901 with measures to curb endemic yellow fever. It is salutary to note that the methods employed were primarily environmental, supplemented by larviciding with oil and some fumigation by sulphur, pyrethrum or burning tobacco waste (Smith 1951). Laird (1967) lists numerous, often successful, examples of the use of fish to reduce infestations of *A. aegypti* in artificial containers in the early days of control. Eradication of the disease, not the mosquito, was the goal, and it was successful because there was no enzootic reservoir on the island. Similar methods essentially rid the Panama Canal Zone and other New World cities of urban yellow fever (Elton 1952) until the discovery of the enzootic "jungle yellow fever" diminished hopes of eradicating the disease. In this enzootic zone *Stegomyia* could not exist without yellow fever. Thus was born the concept of *Aedes aegypti* eradication, already achieved in some Brazilian cities through persistent control efforts (Soper 1967). The story which is well known (Soper 1963, 1965), reached its climax in the mid-1960's with the initiation of the eradication campaign in the United States (Schliessmann 1964).

The failure of the United States to participate earlier was strongly deplored by Soper (1963) who again registered a protest (Soper 1969) against the decision to abandon the U.S. eradication campaign in early 1969. The five years of the original eradication goal had passed, and the end was still not in sight. The impetus of the hemispheric eradication attempt has now been lost as several other nations in addition to the United States have stopped their efforts for political, economic or technical reasons (Reeves 1972a). Brazil, the first country to achieve eradication, became re-infested quite extensively in north-eastern Para (Fraiha 1968), but the situation was not considered difficult to remedy (Scaff and Galvao 1969; Franco 1969). Costa Rica, French Guiana, Honduras, Mexico and El Salvador have become re-infested, while Colombia, Guyana, Surinam, Venezuela and almost all of the Caribbean islands remain infested (Pan American Sanitary Bureau 1971a, 1971b).

The efforts which had begun with the use of oil and fishes became officially organized in 1947 with the help of DDT which was used mainly as a larvicide or, in perifocal spraying, as an adulticide. As is well known, resistance to DDT appeared in 1954 in Trinidad, followed soon by dieldrin resistance (Camargo 1967; Schoof 1967), so that continuing eradication and control attempts at the present time must increasingly rely on the organophosphorus compounds such as Abate, Dursban, fenthion and malathion. Resistance to organophosphorus insecticides, including malathion and Abate, is already known in the Americas, South-east Asia and Africa (World Health Organization 1970).

Outside the Americas there has not even been a moderate-scale eradication or control attempt of *A. aegypti*, perhaps with the exception of Thailand, although most urban areas in Africa and Asia devote some effort to eliminating large infestations and, more significantly, have probably reduced the density of *A. aegypti* through anti malarial premise spraying. In Mauritius, the species was actually eradicated in this unintentional way (Halcrow 1954). In Thailand, Gould et al. (1970, 1971) used a combination of hand-applied larvicide (Abate) and premise fogging with malathion during a three-week period to achieve marked reduction in populations of *A. aegypti*. It is significant that about two months later the treated populations were increasing rapidly at a time when the higher level, untreated populations were declining. This demonstrates the well-known principle that density-dependent factors play an important role in stabilizing populations (Huffaker, Messenger and DeBach, 1971). Trials in a suburb of Bangkok by Kilpatrick, Tonn and Jatanasen (1970) and Lofgren et al. (1970, 1970a) demonstrated that short-term suppression of *A. aegypti* populations could be achieved by ultra-low-volume (ULV) application of malathion by large aircraft. Macdonald (1957) demonstrated that perifocal spraying was both simple and economical for the control of largely domestic populations of *A. aegypti* in Malaya using DDT or dieldrin before resistance to these chemicals had become wide spread. A war-time campaign to control an epidemic of dengue on Guam in 1945 (Bailey and Bohart 1953) apparently resulted in the subsequent disappearance of *A. aegypti* and its replacement by *A. albopictus* which was not known on the island in 1945 (Reeves and Rudnick 1951). In general, however, there has been little attempt at systematic control in South-east Asia or in the Pacific region (Chow 1967; Gratz 1967).

In Africa, control of *A. aegypti* is complicated by the existence of feral populations which have been considered, at least tentatively, as a separate subspecies, *formosus*, distinct from the type form with more domestic or synanthropic habits (Mattingly 1957, 1958, 1967). On a global scale, the two forms may seem largely separate geographically, except in a few areas such as the eastern coast of Africa, but these may be critical to an understanding of the problem. On the basis of this sympatry, interfertility in the laboratory and great variability, I concluded that the species was polymorphic rather than polytypic (McClelland 1967b). Later visits to Newala in southern Tanzania in 1968-70 in three different seasons indicated that the domestic and feral populations were not only distinct in colour characters and behaviour, but that they also appeared to be isolated from each other despite the readiness of the feral populations to bite man and enter houses. The possibility should be considered that the forms of *A. aegypti* may be neither polymorphic nor polytypic but actually more or less isolated incipient species. One may assume that the domestic form arose from the feral form when early man first learned to store water in his house and became better adapted to arid habitats. *A. aegypti* probably is an example of what Bates (1964) has termed "human inquilines". The key factor in the evolution of the synanthropic *A. aegypti* was probably not domesticity *per se* but independence from the natural cycle of rainfall. Once adapted, the domestic form was able to accompany its human host to all the warm parts of the world, whereas the feral form remains confined to sub-Saharan Africa. In Africa, however, the feral form proliferates in the man-made environments of discarded cans, tyres and machinery which receive natural rainfall, occasionally also breeding inside houses where water is stored. Only in those ethnic cultures where water is traditionally stored inside houses does the domestic form occur and reach high levels of infestation (Pichon, Hamon and Mouchet, 1969).

It seems likely that the domestic form competitively excludes the feral form from habitats where water and man are continuously present. The domestic form is, on the other hand, competitively excluded by the feral form in habitats where most breeding sites are filled naturally by rainfall. The ship-borne habitat responsible for spreading the species around the world was in the former category and thus unlikely to transport the feral form effectively. The people of the coastal area of Tanzania around Dar es Salaam, unlike those of Newala, do not store water for long periods in their houses. The domestic form of *A. aegypti* would thus not survive the pressure of the feral form in Dar es Salaam. However, in India (Rao, Pannicker and Reuben, 1970) and in Florida where the feral form is absent, *A. aegypti* breeds in old tyres and tree holes (Porter, Evans and Hughes, 1961) and even leaf axils (Schliessmann 1966) similar to the breeding places of the feral form in Dar-es-Salaam. This seems to explain the apparently anomalous breeding places of the two forms in different areas if habitat is defined by its proximity to houses rather than by the criteria of natural versus artificial water, and when the presence or absence of the competing form is not taken into account.

Recent epidemics of yellow fever in Africa have generally not involved *A. aegypti* except in a few places such as Diorbel, Senegal, where *A. aegypti*, presumably the domestic form, was common (Hamon, Pichon and Cornet, 1971). An epidemic of chikungunya virus disease occurred in 1952 in one of the few areas in East Africa where it is the general custom to store water; the Newala Plateau in southern Tanzania (Lumsden 1955). Domestic populations of *A. aegypti* were abundant in 1953 and even in 1970, despite the generally available piped water. Such socio-culturally determined concentrations of *A. aegypti* are potential epidemic foci, but probably of very localized importance compared with the wide spread sylvatic vectors. It seems clear that the feral populations of *A. aegypti* are rarely involved in epidemics of yellow fever in Africa. This form, the "African *Aedes aegypti*" of Lebrun (1964), is probably not a menace. If it plays little or no part in virus epidemics in Africa, it is unlikely that it would in the Americas even if it succeeded in being introduced and adapting to the different environment. This in itself, as Machado-Allison and Craig (1972) concluded, seems unlikely; it is the paler form that has demonstrated aggressive adaptability in its colonization of America and Asia.

Considerable confusion still exists as to the nature of the difference between the two "subspecies" of *A. aegypti*. Although they are largely allopatric in Africa but cross readily in the laboratory, the hypothesis of subspecificity is seriously challenged in areas like the Newala Plateau mentioned above where domestic populations of the type form are, in the rainy season, surrounded by much larger populations of the feral *formosus*. The confusion has been encouraged by the frequent reference to *formosus* as a forest form with the connotation of lush humid habitats. Clearly in Newala, for example, ambient temperatures and humidities will be comparable in the habitats of both forms. Machado-Allison and Craig (1972) explain the much greater resistance of desiccation of the domestic form as adaptive to the wider range of - and lower - humidities of "open areas" compared to "forests".

Some of the blackest and most extreme examples of the species *formosus* that I have seen came from tree holes in a narrow gallery forest at Garissa in the arid region of north-eastern Kenya where less than 25 cm of rain fell per year in a rainy season of little more than a month. If, however, the ecological cause of the separation of the two forms is dependent on natural rain-filled breeding places versus artificially-maintained water sources, it should then become clear that *formosus* is only present as an adult during the rainy season when, even in the drier climates, humidity is high. Adaptation to resist desiccation is thus unnecessary; if it survived it could not breed. Breeding of the domestic form is largely controlled by human activity. This can result in peak populations in the hottest and driest months of the year (Pichon and Gayral, 1970) when resistance to desiccation in the adult stage would be at a premium. The interesting corollary of this is that the eggs and oviposition behaviour of the two forms might be expected to exhibit a somewhat reverse set of adaptations. This was in fact suggested by the work of Gillett (1955b, 1955c, 1956) whose experiments are now being repeated with sympatric pairs of strains from Newala and elsewhere in Africa. If the two forms of *A. aegypti* were shown to be biological species, hopefully the domestic populations could be easily eradicated in Africa by modification of water-storage habits without fear of their rapid re-establishment from the feral population, except temporarily during the rainy season.

Whereas control of *A. aegypti* in Africa would certainly reduce the hazard of epidemics of yellow fever and perhaps other viruses such as chikungunya, the bulk of cases may continue to arise from sylvatic vectors. Chief among these is *A. simpsoni* which has a characteristic unique among *Stegomyia* but shared with several species of *Anopheles*, *Culex* and *Aedes*, of being an inquiline in man's agro-ecosystem. Just as *Culex tritaeniorhynchus* in Asia or *Aedes nigromaculis* to the irrigated pastures of California, so *A. simpsoni* has adapted to species of *Musa*, *Colocasia* and *Ananas* introduced to Africa from Asia and America. Originally *A. simpsoni* must have been an uncommon species exploiting the specialized niche of water-filled axils of the wild African banana (*Ensete*), and other native species of *Crinum*, *Dracaena*, *Sanseveria*, etc.

At one time,numbers of *A. simpsoni* in the town of Entebbe, Uganda, were effectively reduced by restrictions on the growing of pineapples, *Dracaena* and *Sanseveria* (Gillett 1951 and personal communication). Not all cultivars of *Musa* have suitable axils for the larvae of *A. simpsoni* (Gibbins 1942; Teesdale 1957); therefore suggesting some prospect for environmental control by selective planting. Pineapples and Colocasia, however, would be more difficult (Gillett 1971). In the region of Mount Kilimanjaro, the Chagga people extensively plant a species of *Dracaena* which adds considerably to the production of *A. simpsoni* but, as it has social importance, it could not be abandoned easily although a substitute might be found (Gillett 1969, 1971).

In the Omo River area of Ethiopia, where epidemics of yellow fever occurred in 1960-1962, extensive plantations of *Ensete* supported large populations of *A. simpsoni*, the primarily suspected vector (Neri et al., 1968). This area was later the site of a pilot control project to assess the effectiveness of ULV applications of malathion (Brooks et al., 1970). Biting was almost completely eliminated for about two weeks by a 1,400 gm dose per hectare but only temporarily suppressed by a 420 gm dose. Nevertheless, such partial reduction could be effective in breaking an epidemic, providing control measures could be carried out before the epidemic reached its natural maximum. This was not the first time that aerial spraying against sylvatic yellow fever vectors had been used in Africa. DDT was sprayed at a rate of 200 gm per hectare near Gemena, Zaire, in 1958 and may have contributed to the decline of a yellow fever epidemic (Lebrun 1963).

Like *A. aegypti*, *A. simpsoni* also seems to occur in two behavioural forms but the morphological variation is not clearly correlated (Gillett 1951, 1955a, 1972a,b). Epidemiologically, however, the need for differentiation may be more important as it concerns the degree of man-biting. Both behavioural forms breed in plantations but areas where man-biting occurs do not differ obviously from those where the species only rarely bites man. The earlier hypothesis offered to explain the niche divergence in *A. aegypti* can be extended to *A. simpsoni*. One can assume that *A. simpsoni* became specialized to the plant-axil habitat and was thus a wide spread but not abundant species in Africa until man began to cultivate exotic plants such as bananas, colocasias and pineapples. All of these plants are propagated by offshoots rather than seeds, and thus eggs or larvae of *A. simpsoni* could easily be carried to new plantations in areas where the natural plant hosts were absent. An "agrinquiline" race of *A. simpsoni* could thus have evolved which was closely adapted to breeding in exotic plants and to feeding on man. Where plantations occurred within the original distribution of the "feral" race, either genetic exchange or competitive exclusion might suppress the populations or characteristics of the "agrinquiline" race. In the 1969 epidemic of yellow fever on the Jos Plateau of Nigeria, *A. simpsoni* was breeding in large numbers in axils of native *Crinum* as well as cultivated *Colocasia*, but it was considered non-anthropophilic by Lee and Moore (1972). The hypothesis might be partially tested by comparing the distribution of man-biting and non-man-biting populations of *A. simpsoni* in plantations relative to that of populations breeding in natural vegetation. A better understanding of the apparent behavioural differences between populations of *A. simpsoni* is necessary before control of this important vector can be rationally attempted (Gillett 1971).

The possible competitive relationships between inquiline and feral *A. aegypti* and perhaps *A. simpsoni* may have a counterpart in Asia: in Malaya *A. aegypti*, probably introduced earlier by man, was thought to have displaced to some extent the native species *A. albopictus* from urban areas during the previous 60 years (Macdonald 1958), perhaps through competition (Rudnick 1965). Gilotra, Rozeboom and Bhattacharya (1967) attributed a similar phenomenon in Calcutta to competitive displacement. In more rural areas, *A. albopictus* remained the predominant species (Gilotra and Bhattacharya, 1968). Chan, Chan and Ho (1971), however, in carefully analysing these and earlier observations in the light of their own studies in Singapore, conclude that many of the reports of displacement are erroneous and that, whereas *A. aegypti* is certainly extending its distribution with urban expansion, *A. albopictus* may be itself adapting more to the man-made environment. In a rural area on an island in the Gulf of Thailand, Gould et al. (1968, 1970) found *A. albopictus* to be widely distributed outdoors and more versatile in its breeding habits than *A. aegypti* which was largely restricted to artificial containers and the inside of houses. It was not surprising therefore that treatment of these containers with Abate had little effect on the population of *A. albopictus* but neither did fogging with malathion (Gould et al., 1970).

Apparent competitive displacement between *A. albopictus* and *A. polynesiensis* in laboratory cages (Gubler 1970b; Rozeboom 1971) is perhaps due in part to cross-insemination of *A. polynesiensis* females by *A. albopictus* males (Gubler 1970a) and the apparently similar oviposition habits of the two species (Gubler 1971). These observations led to the initiation of a field trial during which *A. albopictus* was released onto a small Pacific island on which only *A. polynesiensis* had previously been present[1] with the apparent result that *A. albopictus* initially became established without any marked effect on the numbers of *A. polynesiensis*. Whether the population level of *A. polynesiensis* at equilibrium with *A. albopictus* will be lower remains to be seen. It seems unlikely that competitive displacement would have eradicated the native species.

Such attempted ecological manipulations may not be entirely without merit however, the reason being that if the introduced species survives and if both species compete for common resources, it is likely that the population of the native vector species may be reduced (Gillies 1962). The total population of both species could be higher (Ayala 1972) but, if the introduced species is not also a vector, the level of potential transmission of the disease

[1]Field observations on the control of *Aedes polynesiensis* by *Aedes albopictus*. Paper presented at 19th Annual Meeting American Society of Tropical Medicine and Hygiene, San Francisco, November 1970, by L. Rosen and L. Rozeboom.

agent could be lower, and perhaps below the epidemic threshold. In the case of *A. albopictus* and *A. polynesiensis* both, vector potential aside, are severe nuisance biters. From this point of view, the life of the islanders would not be improved. It is possible that diversity increases ecological stability and it might happen that, in a situation with several vector species competing for at least some shared niches, the potential level of transmission might be less than that from a single one of the species alone. If epidemics often result from unusually high vector populations, the unstable conditions leading to such occasional upward oscillations might be rarer and of lower amplitude when more buffered by diversity. The opposite of such enrichment, namely, the often adverse results of creating a vacant niche by the extermination of a pest, has been stressed by Laird (1963) and Macdonald (1972). The dangers of simplification of ecosystems inherent in an intensively controlled monoculture is being recognized today by the development of the pest management concept in agriculture. I would suggest that the application of some of these principles to vector-borne disease control is worthy of more theoretical and practical attention.

Much the same argument applies to the use of both parasites and predators for biological control and in programmes of integrated control (Laird 1963, 1970; Briggs 1970). The permanent establishment of predators or parasites within a pest population results generally not only in lower average absolute levels but also often in a dampening of the oscillations that can carry a vector above the epidemic threshold (Huffaker, Messenger and DeBach, 1971).

The nature of the breeding sites of *Stegomyia* precludes the use of predatory fish (except in artificial containers). The predatory mosquitoes of the genus *Toxorhynchites* have interested many workers and some introductions have probably had some suppressive effect: Paine (1934) estimated only a five per cent reduction of *Aedes scutellaris* in Fiji by *T. splendens*. Nakagawa (1963) admitted that the two species established in Hawaii, *T. splendens* and *T. brevipalpis*, undoubtedly contribute somewhat to the reduction of *A. albopictus* but questioned the value for control. It is interesting, however, that the most colonized habitats were artificial containers, and thus the effect of the predators was at least maximized in the areas of greatest need. The success of these attempts at biological control depended on the establishment of introduced predators. Methods of mass production of *T. brevipalpis* have been developed with a view to controlling *A. aegypti* by inundative releases of a predator which may already occur naturally (Brown 1972).

No account of biological control of *Aedes* would be complete without mention of the Tokelau Islands project (Laird 1967). The parasitic fungus *Coelomomyces stegomyiae* was introduced to larval breeding sites of *A. polynesiensis* on one atoll in 1958. In 1960 the incidence of *C. stegomyiae* was 9 per cent, about fourfold higher than in its native habitats in Singapore. The biting activity of *A. polynesiensis* was about half that observed in 1958. By 1963, the incidence of the fungus was fourfold higher than in 1960.[1] This study was combined with the placement of insecticidal briquettes in larval breeding containers on another atoll, and was thus one of the first examples of integrated control against mosquitos. Recently Laird (1970) has reported that *Coelomomyces* may have disappeared from the atolls and that the results of the project must be considered rather indeterminate. Hamon (1972) suggests that inundative releases of the fungus might have been more effective.

Research proceeds on many possibly pathogenic agents for mosquito control, although with limited resources. Several species are now ready for large-scale field trials (Brown 1972) but none has yet reached the stage of actual operational use in an integrated control programme (Briggs 1970). *Stegomyia* species may pose special problems due to their often cryptic breeding places unless the pathogen is self-inoculative, or is able to disperse with the adult mosquito. Most pathogens, however, are only infective to the larval stages. The difficulty of application pertains generally to all larvicidal agents. Plant axils, in particular, are continually opening but tin cans and other artificial containers are also continually accumulating in the environment. Within a short time after larviciding, new containers will thus be available to ovipositing females and, equally important, a proportion of the toxic containers will become unsuitable or cease to hold water. This was experienced in the Tokelau Project. For these reasons *Toxorhynchites*, among biological control agents for *Stegomyia*, has the advantage that it searches out the cryptic breeding places of its prey yet it is not an obligatory predator. The effort and cost, however, of mass-rearing and inundatively releasing a predator might be better spent in mass-rearing and releasing modified individuals of the target species itself; so-called autocidal control. Although it is possible that mass releases of pathogen-carrying *Stegomyia* females could be used to inoculate breeding places, most methods of autocidal control involve genetic modification.

Rai (1971) lists five categories of genetic control which could be adapted to *Stegomyia*. The well-known sterile male technique has not yet been successfully used against *A. aegypti* though Rai (1969) notes that some of the reasons for the failure of earlier trials can be overcome. The second type of genetic control involves alleged cytoplasmic incompatibility which, while thoroughly studied in *Culex pipiens*, remains little known and conjectural in *A. scutellaris* and not necessarily the same mechanism (McClelland 1967b).

---

[1]Laird, M. (1966) Integrated control and *Aedes polynesiensis:* an outline of the Tokelau Islands Project, and its results, Unpublished mimeograph, World Health Organization/EBL/66.69

A third type of genetic control listed by Rai is interspecific incompatibility but this has been discussed above under competitive displacement. The fourth category involves chromosomal translocations and has been the subject of much work on translocation heterozygotes in *A. aegypti* (Macdonald and Rai, 1970; Rai, McDonald and Asman, 1970; McDonald and Rai, 1971). Fortunately, translocation homozygotes have been obtained recently; their use would greatly increase the potential for control as Whitten (1971) has pointed out in a detailed analysis. Whitten proposed even more effective methods for potential application to mosquitos involving the theoretically possible synthesis of compound chromosomes.

A fifth class of genetic manipulation involves the release of carriers of lethal factors. Among these, the special class of dominant conditional lethal factors perhaps offers the most promise with the added advantage that mass-releases are minimized (Klassen, Knipling and McGuire, 1970; Wehrahahn 1971). Among temperate insects,the most useful traits would probably involve inability to enter diapause or survive low temperatures, but in *Stegomyia* the most likely trait would involve the inability of the eggs to survive a prolonged dry season or a tendency to hatch prematurely. As the hatchability of eggs was one of the first genetic traits to be demonstrated in *A. aegypti* (Gillett 1955c), it gives cause for some optimism.

Another theoretically possible genetic control method involves the use of a meiotic drive mechanism that depresses the ratio of females to males far enough to cause a self-perpetuating population decline (Hickey and Craig, 1966). It might even be possible to use such mechanisms to increase the frequency of a deleterious factor in a natural population (Knipling et al., 1968), particularly conditional lethal factors which do not have the advantage of dominance mentioned above.

None of the new autocidal control methods has been convincingly proven in field trials and most have not progressed beyond the laboratory: most involve research technology that is species-specific. It is probably true that the gap in cost and effort between research and development of a genetic control scheme is greater than that of a chemical insecticide. Once developed, the cost and effort of operational genetic control will probably be no less than that of chemical control.

Many possible methods of biological control and most methods of autocidal control act in a strongly density-dependent manner, rising in efficiency as the population of target insects becomes smaller. Thus, as Whitten (1971) emphasizes, the sterile-male method in particular lends itself to programmes of eradication rather than simple control. The greater capital expense of an eradication programme makes the consequences of recrudescence all the more disastrous. The *Aedes aegypti* eradication campaign has already been discussed; most noteworthy among the autocidal control campaigns is the Southwest Screwworm Eradication Program of the United States. After a low of 153 cases in 1970 and an average of 1,866 cases over the years 1964 to 1971 was achieved, cases in 1972 totalled more than 95,000.[1] The cause was no doubt enhanced by a combination of exceptional weather, inadequate funds and relaxation of the vigilance carried out by cattle producers in pre-control days. One or all of these three factors of weather, cost and complacency, may be expected to influence sooner or later any programme of this nature.

The screwworm campaign, however, has an economically sound cost-benefit ratio of 1:20 (Knipling 1967). There is even a sound economic base for malaria eradication although the campaign is open to serious criticism (Garcia 1972). Apart from emergency epidemic control, how can the massive funds needed for a vector eradication campaign be obtained when the benefit is merely the continued absence of a disease that few people have personally experienced? Such is clearly the case with *A. aegypti* in the Americas and must account in part for the recent setbacks in the campaign.

Thus Sencer (1969) states that there has not been a challenge of the biological efficacy of *A. aegypti* eradication. He lists a number of changed circumstances since the concept of *A. aegypti* eradication was first demonstrated and asks a number of key questions. What is the critical density of the urban vector above which yellow fever virus can reproduce? Can source-reduction and community sanitation keep the density of *A. aegypti* below this critical level? Sencer also makes the comment that species which man has (perhaps unwittingly) eradicated, such as the passenger pigeon, had a marginal ability to survive. On the other hand *A. aegypti* shows an uncanny ability to adapt to new environments. It is a weed-like, colonizing species (McClelland 1967a); this is characteristic of many important pests as well as of man himself (Audy 1968). Bates (1964) states that man has not yet succeeded in eradicating a native pest. Is Reeves (1972a) too pessimistic when he asks: "Can the war to contain infectious diseases be lost?", anticipating another epidemic of classical *A. aegypti*-borne urban yellow fever in his lifetime. Are the dengue epidemics of the Caribbean the early warning of worse, yellow fever, to come? (Reeves 1972b).

There is an enormous gap between much theoretical autocidal control and its application that we confidently say will be filled by more research. I believe that the time has come, not for a frantic increase in better technological tactical weapons against pests

---

[1]Texas A&M University Agricultural Extension Service, Southwest Screwworm Eradication Program, Progress Report for 1972, January 1973, mimeograph.

or *Stegomyia* in particular, but rather for a basic rethinking of goals and strategies.

The first question to resolve must be to decide our priorities in the face of growing competition for available resources. Bruce-Chwatt (1971) states that the diseases that undermine health and efficiency and result in general suffering are often more damaging than the killers, at least to society as a whole. If we accept this, we may decide that our first priority should be to prevent the millions exposed to filarial worms from becoming crippled or blind, rather than to prevent the first handful of deaths that might result from urban yellow fever before surveillance and emergency measures could be effective.

When our priorities are determined, we may realise that most of them are man-made maladies (Audy 1972). In the last few decades, the number of automobiles has greatly increased in developing tropical countries. For every average motor vehicle, there are probably ten or more tyres, many of which will become breeding sites for *A. aegypti* and some of which will be shipped with their mosquito passengers to distant sandal factories. The same may be true of *A. albopictus*, (Eads 1972). Surtees (1971) points to the growing frequency of discarded cans in the tropics, which provide opportunities for increased breeding of *A. aegypti*. The growth of intercontinental airfreight of mechanically-handled palletized cargo would seem to provide far more opportunities for transport of larval mosquitos (Pippin, Thompson and Wilson, 1968) and their successful introduction than would passenger flights. These examples, together with the occurrence of *A. aegypti* in waterpots, the production of *A. simpsoni* on pineapple and banana plantations mentioned above and *Aedes* and *Culex* species in irrigated lands, are surely all examples of externalities (Coale 1970) or the unaccounted costs of much human activity. Just as the best way to conquer the waste disposal problem is to waste less, the recognized method of source reduction is the most logical way of combatting many of the most important mosquito pest problems. Such ecological control is currently practised primarily in those areas where all other methods have failed and the public demands relief (Provost 1969). Management of public health pests, like those of agriculture, may become largely a matter of what Bierne (1971) calls "people management".

It should be remembered that Gorgas broke the cycle of endemic yellow fever by bringing the population of *A. aegypti* below a critical level. This was accomplished without DDT or its later substitutes and without knowledge of the mosquito's genetic mechanism or physiology. The same, or even less, control would be sufficient to prevent an epidemic.

Whereas we must recognize the end of the era of chemical control (Newsom 1970) alone as Gillett (1971) has emphasized, there is a place for the old as well as the new in properly integrated vector control.

This paper was written in part with assistance from Grant No. 1 RO1 AI 10985-01 CDC from the National Institutes of Allergy and Infectious Diseases, NIH. I am grateful to Dr J.D. Gillett and Dr Michael E. Gilpin for their helpful comments on reading the manuscript.

REFERENCES

Audy, J.R. (1968) *Red mites and typhus*, Athlone Press, London, 191 p.

Audy, J.R. (1972) Aspects of human behaviour interfering with vector control. In *Vector control and the recrudescence of vector-borne diseases*, Washington, D.C., 15 June 1971 (Pan American Health Organization, Scientific Publication No. 238), 67-82

Ayala, F.J. (1972) Competition between species, *Am. Scient.*, 60, 348-357

Bailey, S.F. and Bohart, R.M. (1953) A mosquito survey and control program in Guam, *J. econ. Ent.*, 45, 947-952

Bates, M. (1964) Insects in the human ecosystem, *Bull. ent. Soc. Amer.*, 10, 67-70

Bierne, B.P. (1971) Pest management through people management, *Proc. Tall Timbers Conf. Ecol. Anim. Control by Habitat Management*, Tallahassee, Florida, Feb. 25-27, 1971, 3, 1-7

Briggs, J.D. (1970) Principles of integrated control of disease vectors, *Am. Zool.*, 10, 567-571

Brooks, G.D., Neri, P., Gratz, N.G. and Weather, D.B. (1970) Preliminary studies on the use of ultra-low-volume applications of malathion for control of *Aedes simpsoni*, *Bull. Wld Hlth Org.*, 42, 37-54

Brown, A.W.A. (1972) Alternative methods of vector control. In *Vector control and the recrudescence of vector-borne diseases*, Washington, D.C., 15 June 1971 (Pan American Health Organization, Scientific Publication No. 238), 59-63

Bruce-Chwatt, L.J. (1971) Insecticides and the control of vector-borne diseases, *Bull. Wld Hlth Org.*, 44, 419-424

Camargo, S. (1967) History of *Aedes aegypti* eradication in the Americas, *Bull. Wld Hlth Org.*, 36, 602-603

Chow, C.Y. (1967) *Aedes aegypti* in the Western Pacific Region, *Bull. Wld Hlth Org.*, 36, 544-546

Coale, A.J. (1970) Man and his environment, *Science*, 170, 132-136

Eads, R.B. (1972) Recovery of *Aedes albopictus* from used tires shipped to United States ports, *Mosquito News*, 32, 113-114

Elton, N.W. (1952) Yellow fever in Panama: historical and contemporary, *Am. J. trop. Med. Hyg.*, 1, 436-456

Fraiha, H. (1968) Reinfestaçao do Brasil pelo *Aëdes aegypti*. Consideraçoes sôbre o risco de urbanização do virus da febre amarela silvestre na região reinfestada, *Ref. Inst. Med. trop. São Paulo*, 10, 289-294

Franco, O. (1969) Reinfestação do Pará pelo *Aedes aegypti*, *Rev. Bras.Mal. Doenças Trop.*, 21, 729-731

Garcia, R. (1972) The control of malaria, *Environment*, 14, 2-9

Gibbins, E.G. (1942) On the habits and breeding places of *Aedes (Stegomyia) simpsoni* Theobald in Uganda, *Ann. trop. Med. Parasit.*, 36, 151-160

Gillett, J.D. (1951) The habits of the mosquito *Aëdes (Stegomyia) simpsoni* Theobald in relation to the epidemiology of yellow fever in Uganda, *Ann. trop. Med. Parasit.*, 45, 110-121

Gillett, J.D. (1955a) Further studies on the biting behavior of *Aëdes (Stegomyia) simpsoni* Theobald in Uganda, *Ann. trop. Med. Parasit.*, 49, 154-157

Gillett, J.D. (1955b) Variation in the hatching response of *Aëdes* eggs (Diptera: Culicidae), *Bull. ent. Res.*, 46, 241-254

Gillett, J.D. (1955c) The inherited basis of variation in the hatching-response of *Aëdes* eggs (Diptera: Culicidae), *Bull. ent. Res.*, 46, 255-265

Gillett, J.D. (1956) Genetic differences affecting egg-laying in the mosquito *Aëdes (Stegomyia) aegypti* (Linnaeus), *Ann. trop. Med. Parasit.*, 50, 362-374

Gillett, J.D. (1969) *Aedes simpsoni* in Chaggaland, Tanzania, *Ann. trop. Med. Parasit.*, 63, 147-156

Gillett, J.D. (1971) *Mosquitoes*, Weidenfeld and Nicolson, London, 274 p.

Gillett, J.D. (1972a) *Aedes simpsoni* in Chaggaland II - breeding in banana axils, *E. Afr. Med. J.*, 49, 285-290

Gillett, J.D. and Van Someren, E.C.C. (1972b) *Aedes simpsoni* in Chaggaland III - analysis of the biting cycle, *E. Afr. Med. J.*, 49, 291-297

Gillies, M.T. (1962) Interspecific competition in mosquitoes - has it any bearing on biological control? *Proc. 11th Internat. Congr. Ent.*, Vienna, 1960, 11, 502-503

Gilotra, S.K. and Bhattacharya, N.C. (1968) Mosquito vectors of dengue-chikungunya viruses in a rural area near Calcutta, *Bull. Calcutta Sch. Trop. Med.*, 16, 41-42

Gilotra, S.K., Rozeboom, L.E., and Bhattacharya, N.C. (1967) Observations on possible competitive displacement between populations of *Aedes aegypti* Linnaeus and *Aedes albopictus* Skuse in Calcutta, *Bull. Wld Hlth Org.*, 37, 437-446

Gould, D.J., Yuill, T.M., Moussa, M.A., Simasathien, P. and Rutledge, L.C. (1968) An insular outbreak of dengue haemorrhagic fever III. Identification of vectors and observations on vector ecology, *Am. J. trop. Med. Hyg.*, 17, 609-618

Gould, D.J., Mount, G.A., Scanlon, J.E., Ford, H.R. and Sullivan, M.E. (1970) Ecology and control of dengue vectors on an island in the Gulf of Thailand, *J. med. Ent.*, 7, 499-508

Gould, D.J., Mount, G.A., Scanlon, J.E., Sullivan, M.F. and Winter, P.E. (1971) Dengue control on an island in the Gulf of Thailand. I. Results of an *Aedes aegypti* control program, *Am. J. trop. Med. Hyg.*, 20, 705-714

Gratz, N.G. (1967) The control of *Aedes aegypti* in South-east Asia and the Western Pacific, *Bull. Wld Hlth Org.*, 36, 614-617

Gubler, D.J. (1970a) Induced sterility in *Aedes (Stegomyia) polynesiensis* Marks by cross-insemination with *Aedes (Stegomyia) albopictus* Skuse. *J. med. Ent.*, 7, 65-70

Gubler, D.J. (1970b) Competitive displacement of *Aedes (Stegomyia) polynesiensis* Marks by *Aedes (Stegomyia) albopictus* Skuse in laboratory populations, *J. med. Ent.*, 7, 229-235

Gubler, D.J. (1971) Studies on the comparative oviposition behavior of *Aedes (Stegomyia) albopictus* Skuse and *Aedes (Stegomyia) polynesiensis* Marks. *J. med. Ent.*, 8, 675-682

Halcrow, J.G. (1954) Catalogue of the mosquitoes of Mauritius and Rodrigues, *Bull. Mauritius Inst.*, 3, 234-248

Hamon, J. (1972) The response of vectors and of nature to current eradication and control measures. In *Vector control and the recrudescence of vector-borne diseases*, Washington, D.C., 15 June 1971 (Pan American Health Organization, Scientific Publication No. 238), 54-58

Hamon, J., Pichon, G. and Cornet, M. (1971) La transmission du virus amaril en Afrique occidentale. Ecologie, répartition, fréquence et controle des vecteurs, et observations concernant l'épidémiologie de la fièvre jaune, *Cah. O.R.S.T.O.M., Sér. Ent. méd. Parasit.*, 9, 3-60

Hickey, W.A. and Craig, G.B. Jr. (1966) Distortion of sex ratio in populations of *Aedes aegypti*, *Can. J. Genet. Cytol.*, 8, 260-278

Huffaker, C.B., Messenger, P.S. and DeBach, P. (1971) The natural enemy component in natural control and the theory of biological control. In Huffaker, C.B. (ed) *Biological Control*, Plenum Press, New York, 17-67

Kilpatrick, J.W., Tonn, R.J. and Jatanasen, S. (1970) Evaluation of ultra-low-volume insecticide dispensing systems for use in single-engine aircraft and their effectiveness against *Aedes aegypti* populations in South-east Asia, *Bull. Wld Hlth Org.*, 42, 1-14

Klassen, W., Knipling, E.F. and McGuire, J.W. Jr. (1970) The potential for insect-population suppression by dominant conditional lethal traits, *Ann. ent. Soc. Am.*, 63, 238-255

Knipling, E.F. (1967) Sterile male technique - principles involved, current application, limitations, and future application. In *Genetics of insect vectors of disease*, Wright, J.W. and Pal, R. (eds) Elsevier, Amsterdam, 587-616

Knipling, E.F. *et al.* (1968) Genetic control of insects of public health importance, *Bull. Wld Hlth Org.*, 38, 421-438

Laird, M. (1963) Vector ecology and integrated control procedures, *Bull Wld Hlth Org.*, 29 (Suppl.), 147-151

Laird, M. (1967) Biotic factors in the control of *Aedes aegypti*, *Bull.Wld Hlth Org.*, 36, 625-627

Laird, M. (1970) Integrated control of mosquitoes, *Am. Zool.*, 10, 573-578

Lebrun, A.J. (1963) Jungle yellow fever and its control in Gemena, Belgian Congo, *Am. J. trop. Med. Hyg.*, 12, 398-407

Lebrun, A.J. (1964) The African *Aedes aegypti*, *Mosquito News*, 24, 389-393

Lee, V.H. and Moore, D.L. (1972) Vectors of the 1969 yellow fever epidemic on the Jos Plateau, Nigeria, *Bull. Wld Hlth Org.*, 46, 669-673

Lofgren, C.S., Ford, H.R., Tonn, R.J. and Jatansen, S. (1970) The effectiveness of ultra-low-volume applications of malathion at a rate of 6 US fluid ounces per acre in controlling *Aedes aegypti* in a large scale test at Nakhon Sawan, Thailand, *Bull. Wld Hlth Org.*, 42, 15-25

Lofgren, C.S., Ford, H.R., Tonn, R.J., Bang, Y.H. and Siri-Bodhi, P. (1970a) The effectiveness of ultra-low-volume applications of malathion at a rate of 3 US fluid ounces per acre in controlling *Aedes aegypti* in Thailand, *Bull. Wld Hlth Org.*, 42, 27-35

Lumsden, W.H.R. (1955) An epidemic of virus disease in Southern Province, Tanganyika Territory, in 1952-53, *Trans. R. Soc. trop. Med. Hyg.*, 49, 33-57

Macdonald, W.W. (1957) *Aëdes aegypti* in Malaya III. Experiments on control and eradication, *Ann. trop. Med. Parasit.*, 51, 175-186

Macdonald, W.W. (1958) Some ecological factors influencing the distribution of Malayan mosquitoes, *Proc. Centenary and Bicentenary Congress of Biology* (Singapore), 117-122

Macdonald, W.W. (1972) The response of vectors and of nature to current eradication and control measures. In *Vector control and the recrudescence of vector-borne disease*, Washington, D.C., 15 June 1971 (Pan American Health Organization, Scientific Publication No. 238), 41-53

Machado-Allison, C.E. and Craig, G.B. (1972) Geographic variation in resistance to desiccation in *Aedes aegypti* and *A. atropalpus* (Diptera: Culicidae), *Ann. ent. Soc. Am.*, 65, 542-547

Mattingly, P.F. (1957) Genetical aspects of the *Aëdes aegypti* problem I. Taxonomy and bionomics, *Ann. trop. Med. Parasit.*, 51, 392-408

Mattingly, P.F. (1958) Genetical aspects of the *Aëdes aegypti* problem II. Disease relationships, genetics and control, *Ann. trop. Med. Parasit.*, 52, 5-17

Mattingly, P.F. (1967) Taxonomy of *Aedes aegypti* and related species, *Bull. Wld Hlth Org.*, 36, 552-554

McClelland, G.A.H. (1967a) Problems of interaction and density in mosquitoes in relation to *Aedes aegypti* eradication, *Proc. 35th Ann. Conf. Calif. Mosquito Control Assoc.*, February 1967, 94-95

McClelland, G.A.H. (1967b) Speciation evolution in *Aedes*. In *Genetics of insect vectors of disease*, Wright, J.W. and Pal, R. (eds) Elsevier, Amsterdam, 277-311

McDonald, P.T. and Rai, K.S. (1970) *Aedes aegypti*: origin of a new chromosome from a double translocation heterozygote, *Science*, 168, 1229-1230

McDonald, P.T. and Rai, K.S. (1971) Population control potential of heterozygous translocations as determined by computer simulations, *Bull. Wld Hlth Org.*, 44, 829-845

Nakagawa, P.Y. (1963) Status of *Toxorhynchites* in Hawaii, *Proc. Hawaiian Ent. Soc.*, 18, 291-293

Neri, P., Série, C., Andral, L. and Poirier, A. (1968) Etudes sur la fièvre jaune en Ethiopie. 4. Recherches entomologiques à la station de Manéra, *Bull. Wld Hlth Org.*, 38, 863-872

Newsom, L.D. (1970) The end of an era and future prospects of insect control, *Proc. Tall Timbers Conf. Ecol. Anim. Control by Habitat Management*, Tallahassee, Florida, Feb. 26-28, 1970, 2, 117-136

Paine, R.W. (1934) The introduction of *Megarhinus* mosquitoes into Fiji, *Bull. ent. Res.*, 25, 1-32

Pan American Sanitary Bureau (1971a) Informe de Marzo de 1971 sobre la campaña de eradication del *Aedes aegypti* en las Americas, *Bol. Of. San. Panam.*, 70, 586

Pan American Sanitary Bureau (1971b) Reinfestaciones por *Aedes aegypti* en Costa Rica, *Bol. Of. San. Panam.*, 71, 230

Pichon, G., Hamon, J. and Mouchet, J. (1969) Groupes ethniques et foyers potentiels de fièvre jaune dans les Etats francophones d'Afrique occidentale; considérations sur les méthodes de lutte contre *Aedes aegypti*, *Cah. O.R.S.T.O.M.*, *Sér. Ent. méd. Parasit.*, 7, 39-50

Pichon, G. and Gayral, P. (1970) Dynamique des populations d'*Aedes aegypti* dans trois villages de savane d'Afrique de l'Ouest. Fluctuations saisonnières et incidence épidémiologique, *Cah. O.R.S.T.O.M. Sér. Ent. méd. Parasit.*, 8, 49-68

Pippin, W.F., Thompson, S. and Wilson, R. (1968) The interception of living larvae of *Aedes aegypti* (L.) and *Culex cinerellus* Edw. in aircraft, *Mosquito News*, 28, 646

Porter, J.R., Evans, B.R. and Hughes, J.H. (1961) The significance of trees as mosquito foci, *Mosquito News*, 21, 234-237

Provost, M.W. (1969) Ecological control of salt marsh mosquitoes with side benefits to birds, *Proc. Tall Timbers Conf. Ecol. Anim. Control by Habitat Management*, Tallahassee, Florida, Feb. 27-28, 1969, 1, 193-206

Rai, K.S. (1969) Status of the sterile-male technique for mosquito control. In *Sterile-male technique for eradication or control of harmful insects*. Doyle, E. and Freeman, S.M. (eds), Internat. Atomic Energy Agency, Vienna, 107-114

Rai, K.S. (1971) The prospect for genetic control of filariasis vectors, *Jap. J. Genet.*, 46, 207-214

Rai, K.S., McDonald, P.T. and Asman, M. (1970) Cytogenetics of two radiation-induced, sex-linked translocations in the yellow fever mosquito, *Aedes aegypti*, *Genetics*, 66, 635-651

Rao, R.T., Pannicker, K.N. and Reuben, R. (1970) Tree-hole breeding of *Aedes aegypti* in southern India: a preliminary report, *Bull. Wld Hlth Org.*, 42, 333-334

Reeves, W.C. (1972a) Can the war to contain infectious diseases be lost? *Am. J. trop. Med. Hyg.*, 21, 251-259

Reeves, W.C. (1972b) Recrudescence of arthropod-borne virus diseases in the Americas. In *Vector control and the recrudescence of vector-borne diseases*, Washington, D.C., 15 June, 1971 (Pan American Health Organization, Scientific Publication No. 238 3-14)

Reeves, W.C. and Rudnick, A. (1951) A survey of the mosquitoes of Guam in two periods in 1948 and 1949 and its epidemiological implications, *Am. J. trop. Med.*, 31, 633-658

Rozeboom, L.E. (1971) Relative densities of freely breeding populations of *Aedes (S.) polynesiensis* Marks and *A. (S.) albopictus* Skuse. A large cage experiment, *Am. J. trop. Med. Hyg.*, 20, 356-362

Rudnick, A. (1965) Studies of the ecology of dengue in Malaysia: a preliminary report, *J. med. Ent.*, 2, 203-208

Scaff, L.M. and Galvao, S. dos S. (1969) A reinfestação do Pará pelo *Aedes aegypti* situação atual, *Rev. Bras. Mal. Doenças Trop.*, 21, 615-626

Schliessmann, D.J. (1964) The *Aedes aegypti* eradication program of the U.S., *Mosquito News*, 24, 124-132

Schliessmann, D.J. (1966), Progress report of the *Aedes aegypti* eradication program in the United States for 1965, *Mosquito News*, 26, 486-489

Schoof, H.F. (1967) Insecticides for use against *Aedes aegypti*, *Bull. Wld Hlth Org.*, 36, 618-622

Sencer, D.J. (1969) Health protection in a shrinking world, *Am. J. trop. Med. Hyg.*, 18, 341-345

Smith, H.H. (1951) Controlling yellow fever. In *Yellow Fever*, Strode, G.K. (ed), McGraw Hill, New York, 539-628

Soper, F.L. (1963) The elimination of urban yellow fever in the Americas through the eradication of *Aedes aegypti*, *Am. J. Pub. Hlth.*, 53, 7-16

Soper, F.L. (1965) The 1964 status of *Aedes aegypti* eradication and yellow fever in the Americas, *Am. J. trop. Med. Hyg.*, 14, 887-891

Soper, F.L. (1967) *Aedes aegypti* and yellow fever, *Bull. Wld Hlth Org.*, 36, 521-527

Soper, F.L. (1969) Correspondence, *Am. J. trop. Med. Hyg.*, 18, 482-484

Surtees, G. (1971) Urbanization and the epidemiology of mosquito-borne disease, *Abstr. Hyg.*, 46, 121-134

Teesdale, C. (1957) The genus *Musa* Linn. and its role in the breeding of *Aedes (Stegomyia) simpsoni* (Theo.) on the Kenya coast, *Bull. ent. Res.*, 48, 251-260

Wehrhahn, C.F. and Klassen, W. (1971) Genetic insect control methods involving the release of relatively few laboratory reared insects, *Can. Ent.*, 103, 1387-1396

Whitten, M.J. (1971) Insect control by genetic manipulation of natural populations, *Science*, 171, 682-684

World Health Organization (1970) Insecticide resistance and vector control, *Wld Hlth Org. techn. Rep. Ser.* No. 443, 279 p.

Additional reference:

Chan, K.L., Chan, Y.C. and Ho, B.C. (1971) *Aedes aegypti* (L.) and *Aedes albopictus* (Skuse) in Singapore City. 4. Competition between species, *Bull. Wld Hlth Org.*, 44, 643-649

DISCUSSION OF THE PAPER ON *STEGOMYIA*

A.W.A. Brown

Vector Biology and Control

World Health Organization

Dr McClelland, in summarizing the achievements and prospects in the control of *Stegomyia* mosquitos, has told us much about the yellow-fever vector *Aedes simpsoni*, but quite rightly has paid very great attention to *Aedes aegypti*, the urban vector of yellow fever and of the dengue viruses. With respect to *A. aegypti* he has told us that much of its control or management depends on "people management". With respect to chemical control, he cites the experience of Gould and co-workers in Thailand that chemically treated populations subsequently recovered to higher numbers than before, and quotes Newsom's conclusion that we must recognize the end of the era of chemical control.

The present day distribution of urban *A. aegypti* is largely a consequence of the human living conditions in different parts of the world. The disappearance of this species from the entire Mediterranean area, in contrast to its persistence in the Caribbean islands and mainland, and its reappearance in Central America and Brazil, is due to the improvements in piped water and sanitation and often the use of DDT, in a comparatively northern outlier of this tropical species. Nevertheless, control operations have eliminated it from many Red Sea ports; it is rare in the Persian Gulf and Iran, and only moderately distributed in Pakistan and western India. In the Pacific, it is no longer found in Japan, is rare in Taiwan and disappearing from continental Queensland.

In East Africa, *A. aegypti* is abundant only in districts such as Newala in southeast Tanzania or in certain villages north-east of Mombasa, Kenya. In West Africa, the incidence is moderate in most towns and villages, seldom exceeding a house index of 25, although high enough to be receptive to yellow-fever transmission in Senegal, northern Ghana, Upper Volta and parts of Nigeria.

In South-east Asia, the highest densities of *A. aegypti* are found. House indices frequently exceed 95 per cent and Breteau indices occasionally exceed 500 positive containers per 100 houses. Jars of stored water and the ant-traps at the base of table legs are the most common larval sites. It is in these countries, notably Burma, Thailand, Vietnam and the Philippines, that epidemics of dengue haemorrhagic fever (DHF) continuously occur, transmitted to young children in the family by *A. aegypti* Only in Malaysia are the indices moderate, whereas in Singapore they are considerably reduced.

All of this information, based on special surveys and on information in the literature, has been collected and stored in the computer at WHO Headquarters, Geneva. In order to transpose the information onto maps, conversion factors have been derived to allow house indices, container indices or Breteau indices to be converted to single digit figures in a series from 1 to 9. These are printed out on 34 maps of different parts of the Old World on a scale of one to five million, and have been distributed on a scale of one to eight million; they are available on request on the larger scale.

Dr McClelland must have been much gratified that Singapore achieved its remarkable control of urban *A. aegypti* by "people management", brought about by the Destruction of Disease-Bearing Insects Act of 1968, which had been preceded by a campaign of publicity and education. Any householder in an area of DHF transmission found breeding *A. aegypti* on his premises is served an order to destroy them; the penalty for failure to comply is a fine of from 20 to 100 Singapore dollars. This method alone was found to reduce the house indices from approximately 9 per cent to approximately 3 per cent. The remainder of the breeding persists in essential household containers, although refractory remainder cases did persist in a recent clean-up campaign in Maracay, Venezuela. Among these, the ant-trap may be treated with a teaspoon of common salt. The large jars in which drinking water is stored, particularly characteristic of Thailand, can be freed of *A. aegypti* larvae by treating them with one per cent Abate in sand granules, added to give an Abate concentration of 1 ppm. A 100-litre jar thus treated with 10 gm sand granules remains limpidly clear and free of *A. aegypti* for the subsequent two and a half months. This method, developed at WHO *Aedes* Research Unit in Bangkok, provides year-round protection for drinking water (and ant-traps) from four treatments per year. It is entirely safe as Abate is virtually non-poisonous to vertebrates regardless of the dosage.

Only a few of the areas with dengue haemorrhagic or yellow fever are interested in this thorough method of prevention. Therefore the *Aedes* Research Unit has also developed chemical control measures for corrective or emergency action if epidemics occur. Thermally-generated fogs of malathion dispersed in an oil solution from generators such as the Swing-Fog, Dyna-Fog and TIFA - not uncommon machines in tropical countries - give control of the adult mosquito population immediately, but they are back within a few days. Thus, attention has been paid to the non-thermal dispersion of concentrated malathion or fenitrothion by the ULV (ultra-low volume) method of application; here the droplet size is about 20 microns compared to 8 microns in fogs, and is sufficient to provide a residual effect. A ULV machine that can be mounted on a trailer and driven along the streets of a town is the LECO generator. With three successive applications of malathion at six ounces per acre, adult control continued for 12 days thereafter; with five applications of fenitrothion at six ounces per acre, the control of *A. aegypti* continued for a subsequent five months.

Excellent results were also obtained by the *Aedes* Research Unit with a hand-portable ULV generator known as the Mitz Moe emitting fenitrothion at 24 ounces per acre. In this case, the rooms of the houses were treated thoroughly, and with two applications, the second after an interval of a fortnight, complete control was sustained for at least four months.

Finally, tests were made with small Cessna or large DC-3 aircraft to be used in the event of an emergency to spray villages, towns or cities. ULV malathion was dispersed in a fine mist-spray at six ounces per acre. With a single application, complete initial control was obtained but the population returned to normal after one week. However, with two applications four days apart, the population of *A. aegypti* was still only one-tenth of the normal ten days later.

Before the development of methods of genetic control with perhaps translocation strains, or of methods of biological control such as *Toxorhynchites* predators in outdoor situations, chemical control methods with safe degradable organophosphorus insecticides are available to those that wish to take advantage of them to prevent or abate epidemics of *Stegomyia*-borne arboviruses.

TSETSE FLIES

K. Hocking

WHO Consultant

INTRODUCTION

It is probably no exaggeration to state that the future of tropical Africa depends to a large extent on the efficient control of the tsetse fly (*Glossina* spp), the carrier of trypanosomiasis. There is a belt centered on the equator, which covers over four million square miles (over ten million square kilometres) of the African continent in which tsetse flies prevent the existence of cattle and horses. In many parts, they are responsible for diseases of man as well. Tsetse flies are important only as vectors of diseases, human and animal. Hence, the only reason for controlling them is to reduce or to eliminate these diseases in order to enable man and cattle to occupy land previously held by some species of *Glossina*.

Generally, the diseases can be prevented or cured by the administration of drugs. Repeated surveys and constant surveillance with mass treatment have reduced the incidence of human trypanosomiasis to a very low level. Treatment of the disease is readily available and highly effective; a large proportion of such infections can be cured and the mortality rate from human sleeping sickness has been very greatly diminished. The situation with regard to animal trypanosomiasis (Nagana) has also improved. Curative treatment with trypanocidal drugs has been increasingly successful; a degree of protection against heavy infection and high mortality is conferred, especially when the fly density and fly-infection rate are relatively low. However, it appears that the risks of drug resistance are greater and tend to complicate the problem of control by chemotherapy, if herds of cattle, for instance, are exposed for unduly long periods.

Eradication of the disease by chemotherapeutic methods has not proved possible. Therefore, for a permanent effect, it is necessary to eradicate the vector. In many cases, a reduction in the numbers of the vectors, or a reduction in the contact between man and fly or cattle and fly is the cheapest and quickest way of controlling these diseases.

THE PRESENT POSITION REGARDING CONTROL METHODS

Indirect Methods

Until the availability of persistent insecticides such as DDT in 1945, all practical methods of control were based on indirect attack on the fly by the removal of vegetation or the destruction of game.

Game destruction. As tsetse flies are entirely dependent on mammalian (or reptilian) blood for their food supply, it is obvious that if all animals are removed from a certain area, tsetse flies cannot survive. This has served as the basis for control methods, particularly in Rhodesia and Uganda. Although this method is still used to a limited extent, it has become very much more selective. Now that there is much information on the preferred hosts of many species of tsetse fly, chiefly from the serological identification of blood meals, attempts are being made to remove only the preferred host animals. However, with the realization that big game is a considerable economic asset to a country and with pressure from conservationists, it is unlikely that control by game destruction will continue much longer.

Removal of vegetation. The clearing of vegetation that entails the removal of all woody growth in a fly-infested area of land is not normally employed as a tsetse-fly control measure, although there are instances of extensive stretches having been denuded for this purpose. It has been and still is customary to clear all the shrubs and trees from selected strips of land in order to check tsetse fly encroachment in conjunction with other methods of control or to reduce the contact between fly and man at river crossings and along the main routes of communication and cattle movement. One of the major disadvantages of these ruthlessly cleared barriers is that they are costly and must be maintained in order to remain effective against re-invasion for a considerable period.

The more useful method of bush-clearing for tsetse fly control is partial - discriminative or selective clearing. An example of which is the removal of vegetation from "interzones" favoured by the fly or of a particular type of vegetation such as thickets on which the species depends. In some areas, it has been found necessary to clear the vegetation from drainage lines and to thin out the surrounding mixed woodland to a tree density unsuitable to the fly population. It is probable that these methods of tsetse control will continue to become more refined as the knowledge of the ecology of the various species increases. Human settlement of the bush with a low level of fly infestation leads to the gradual destruction of the bush and the eventual disappearance of the fly.

Direct Methods

Use of residual deposits of insecticides. The most firmly established of all methods of insecticidal control of tsetse flies is the application of residual deposits to some part of the insects' habitat. This technique was first used against *G. fuscipes* and a few years later against the closely related, riverine species *G. palpalis* and *G. tachinoides*. The extension of the method to the wider ranging savanna species awaited detailed information on their resting sites and dry-season refuges. The details of the application depend very much on the species concerned and the circumstances in which it lives, but the principle is the same in all cases: to apply a relatively heavy and persistent deposit to selected parts of a fly's habitat. No attempt is made to spray all the places which a fly may visit but only to make a sufficiently high proportion of them lethal. All suitable substrates in a limited part of the habitat may be treated or only a selection of them. The appropriate resting sites must remain lethal for a sufficiently long period to allow the last pupa that was in the ground at the start of the operation to mature into an adult; if the kill has been complete, that is sufficient. Residual deposits usually lose their effectiveness slowly. Therefore, there is a considerable safety margin in the life of the deposit after it is no longer completely effective; it is capable of eliminating the offspring of any flies that escape shortly after the initial treatment. In fact, single applications of modern insecticides are very effective in the beginning and apparently provide complete eradication in a limited area if properly applied. The appropriate insecticide and formulation depend very largely on the climate. When the dry season is very harsh and prolonged, DDT wettable powders are adequate. If the deposit has to withstand rainfall, emulsion concentrates are necessary: deposits from dieldrin emulsifiable concentrate remain lethal for a longer period than those from DDT emulsifiable concentrate. Burnett, Robinson and Leroux (1957) in Kenya found that on leaves DDT lost some 60 per cent of its original 0.3 gm/m^2 in five days and most of its lethality, whereas dieldrin at 0.45 gm/m^2 fell by 60 per cent in about 50 days and remained highly lethal for at least eight weeks. Hocking (1961) observed that, on bark, an initial deposit of 0.8 gm dieldrin per square metre lost approximately 60 per cent in 23 days and 90 per cent in four months, but was still lethal.

In field trials, the treatment is based on information obtained from field surveys. If no flies are discovered after the end of the first pupal period, the search is intensified in an attempt to find isolated survivors or immigrant flies. If these are found, a limited second application is usually made in the places where they are discovered. Flies may not be present in detectable numbers until after the next favourable breeding season during which time, the residual population will have had an opportunity to increase. In general, as experience is gained and also as the area under treatment is enlarged, the need for repeated applications diminishes. Many workers have commented on the spectacular decline in tsetse numbers as soon as an insecticide is applied. It is likely to be characteristic of insecticide treatment in general, that if the decline is less than spectacular, the final effect may be unsatisfactory. Therefore, repeated applications would be necessary in order to achieve eradication.

Residual treatments are sometimes used without intending to achieve complete extermination, as, for example, to arrest trypanosomiasis epidemics, to protect specific sites or to permit settlement. It is then unnecessary to cut protective barriers and less complete coverage is required; thus considerable economies are achieved.

Application of insecticides by aircraft. Many species of tsetse flies occur over wide areas, often many hundreds of square miles. In order to control such species, the use of aircraft for the application of insecticide has obvious advantages - their ability to cover large areas quickly.

Application from the air has been found to be uneconomical, even if it is possible, against species living in high forest, but savanna woodland with evergreen thickets can be successfully treated. The only effective treatments have employed the insecticide in the form of a coarse aerosol containing as much as possible of its volume in droplets between 5 and 50 microns in diameter. Larger droplets are unable to bypass obstacles such as leaves and branches to penetrate the habitat of the tsetse flies; smaller droplets do not deposit very effectively on the insects themselves. In this connection, it is of interest to note that the diameters of the single droplets which contain the L.D. 95 for *Glossina morsitans* are as follows:

	Dieldrin 20%	Endosulfan 20%	Isobenzan 20%	Fenthion 95%
for male flies	40 μ	30 μ	27 μ	34 μ
for pregnant females	80 μ	62 μ	56 μ	34 μ

(Calculated from Burnett, 1961-62). The first three insecticides being solids have to be dissolved in solvents before they can be used in this way, whereas fenthion is already a liquid. Under tropical conditions, applications of coarse aerosols are effective only for a limited period (one or two hours) just after dawn and, unreliably, for an even shorter period before dusk. During the day, meteorological conditions are so unstable that rapid diffusion of the aerosol occurs, even before penetrating the bush canopy.

Aerosols have been produced from aircraft by emitting an oil solution of insecticide under high pressure through an extended boom, by injection of the solution into the exhaust of the aircraft and by various types of rotary atomizers. It has not been possible, by the first method, to reduce the mean mass diameter of the droplets in the aerosol below 70 microns and even this is possible only with a reasonably fast aircraft, as speed is needed to assist the break-up of the liquid. An aerosol of approximately the same droplet spectrum can be produced by a light aircraft by the "micronair" equipment produced by Britten-Norman, Ltd., which can be modified to produce aerosols in the 10 to 40 micron range provided the insecticide is in a volatile solvent such as kerosene. Although this technique relies on an appreciable degree of solvent evaporation, there is a negligible loss of the insecticide itself. Injection into the exhaust system can readily produce droplets of the desired size. However, it involves some decomposition of the insecticide by heat, but the use of a suitable solvent and care in the selection of the point of injection can reduce this to a minimum. The volume dosage applied varies from 0.004 gallons to 0.025 gallons per acre (40-250 ccs per hectare), the amount of insecticide from 0.1 to 0.5 ounces per acre (5-6 grams per hectare).

When spraying an area of woodland, an aircraft is flown along parallel runs at regularly spaced intervals (generally 50-70 yards). The aircraft flies as close as possible to the tops of the trees, preferably at right angles to the prevailing wind. There is only a nominal overlap of areas treated on successive days; but treatments of a whole block of woodland are repeated to cover approximately two pupal periods, usually six applications at three-week intervals.

The cost of six treatments depends on the site of operations (distance from supplies, airfield, etc.) and the insecticide and aircraft used. However, the cost cannot be much less than £200 per square mile (US $\$200/km^2$).

Ground disseminated aerosols and smokes. Machines are available to produce a cloud which drifts with the wind and has no residual effect. It resembles that produced by aircraft, but it is already below the canopy and has to be kept airborne. The droplet size should therefore be less than for aerosols disseminated by aircraft. As there is little or no residual effect, applications have to be repeated as mentioned above.

Both the Swing-fog pulse jet machine and the Todd Insecticide Fog Applicator have been used to eliminate tsetse species from pockets of infested bush of up to 30 square miles as in Zambia. All fog generators, however, suffer from low output. The drop size regulator is invariably a flow-rate control; the maximum output is obtainable only with the largest droplets, although this is not apparent from the manufacturers' instructions. The work in Zambia used droplets of from 90 to 120 microns. This is far too large for effective cover in light winds and it is unlikely that many of those drops drifted more than a few yards.

FUTURE PROSPECTS

It is probable that in the immediate future, there will be more and more emphasis on the control of tsetse flies by the selective application of residual insecticides. This is, at present, the most effective method and the easiest to apply. As knowledge of the habits and various species of *Glossina* grows, selective applications can be made more precise, more economical and more effective. Considerable improvements can also be made in the equipment used for applying the very effective insecticides now available. Determining precisely where the insecticide should be deposited is dependent on the availability of equipment that can do so without enormous wastage. Efforts are being made to find insecticides to replace DDT and dieldrin. However, as of the present, others do not persist long enough to replace these materials economically.

Ultra-low-volume techniques which use undiluted insecticides such as malathion or highly insecticidal concentrates, now well-established in other fields, will be extended to applying ground-generated aerosols for tsetse control. If suitable equipment is developed for use in the African bush, this method has great prospects for the future.

Because of the low reproductive potential of tsetse flies and because of the desirability in many cases of eliminating the last fly, the sterile-male genetic control method is very attractive, theoretically. The problem of breeding large numbers of flies in the laboratory appears to have been largely overcome, at least for some species (Nash, Jordan and Boyle, 1968) and methods of sterilizing males are available although not yet without some loss of competitiveness. Other genetic methods such as chromosome translocations (Curtis 1968) are being investigated. With this technique, both sexes can be used, therefore a smaller breeding colony is required.

As sterile-male eradication techniques depend upon a preponderance of introduced males over wild males, perhaps it would be advisable to reduce the resident adult population to a very low level by one application of insecticide (preferably from the air), before the introduction of sterile males. It is probable that this would be more economical and it would certainly give results much more quickly. Eradication of approximately 100 per cent of the male fly population would not prove too difficult as they are many times more susceptible to insecticides such as dieldrin than are the females (particularly pregnant females). Introduced sterile males would then have to compete only with the newly emerging males, which would at first be very few and then would rapidly decrease.

Another approach which deserves more attention is to use aerial applications of insecticide to reduce the tsetse population to a low level, which can be done much more cheaply than complete eradication and to complete the process by settling the area concerned with the necessary density of farmers. If eradication is not required, the dosage by aerial application can be reduced to a minimum and the swathe width can be increased, thus reducing the amount of flying. In this way, only three or four treatments are required. In many areas of Africa where there is pressure on the land, this approach is the obvious one. However, any control method should be considered to be an integral part of the overall development plan for the area concerned.

RESEARCH REQUIREMENTS

Considerably more study of the general ecology of the various species of tsetse fly needs to be carried out, particularly in relation to control by the selective application of insecticides. There is no doubt that more precise knowledge of the habits of any species, at different seasons of the year in a particular habitat, would enable at least ten-fold reductions to be made in the amount of insecticide that had to be used to eradicate that species. Apart from the economic aspects, the smaller the area of the habitat that is treated with persistent insecticides, the less effect there will be on the general ecology of the area. The importance of this effect is, at present, largely unexplored.

Another area of research in which a great deal has been done but which is still far from satisfactorily resolved, concerns methods of assessing the results of control operations, that is, determining population densities at very low levels. Traps and attractants would seem to be able to solve this problem. However, nothing very satisfactory has emerged as yet.

There is also a need to develop more satisfactory equipment, both for applying insecticides in a precise manner to the right places for residual action and for producing aerosols of the right drop spectrum, at a high enough emission rate, from ground operated machines, which can be used for ultra-low-volume applications in the African bush.

It is also desirable to find insecticides to replace the organochlorine compounds at present in use, as one day they may no longer be available. Owing to the length of time that persistence is required to make the residual applications effective, a special problem arises. It may be possible to produce something which persists for a long time on the bark of a tree but which decomposes when washed down into the soil. For aerosols either for aerial or ground application, persistence is not required, and it should not be impossible to find an insecticide which is as effective and as economical to produce as those presently used, but which is rapidly bio-degradable. If it is also, like fenthion, as effective against pregnant females as against males, so much the better. One of the synthetic pyrethroids, now being produced in large numbers, may prove to be the answer.

SUMMARY

Originally indirect methods for the control of tsetse flies by removing their vegetation cover or their food supply were used. Game destruction even though it has become more selective is on its way out. Certain aspects of bush clearing will always have their place in control operations.

With the advent of cheap and highly effective insecticides, the chief method of control has become the selective application of residual deposits of insecticides, although the application of insecticidal aerosols from aircraft is valuable in certain circumstances, and can be comparable in cost.

For the immediate future, refinement of the selective application of insecticides is the most promising line but genetic methods of control will probably be used in the not too distant future to control this sparse and slow-breeding insect.

REFERENCES

Burnett, G.F., Robinson, J. and Le Roux, J.G. (1957) Comparative trials of DDT and dieldrin for the control of the riverside tsetse *Glossina palpalis fuscipes* (Newst.), *E. Afr. agric. J.*, 22, 142-146

Burnett, G.F. (1961) The susceptibility of tsetse flies to topical applications of insecticides. I. Young adults of *Glossina morsitans* and chlorinated hydrocarbons, *Bull. ent. Res.*, 52, 531-539

Burnett, G.F. (1962) The susceptibility of tsetse flies to topical applications of insecticides. VI. Data on more chlorinated hydrocarbons and organophosphates and a general discussion, *Bull. ent. Res.*, 53, 753-761

Curtis, C.F. (1968) A possible genetic method for the control of insect pests, with special reference to tsetse flies (*Glossina* spp.), *Bull. ent. Res.*, 57, 509-523

Hocking, K.S. (1961) Discriminative application of insecticide against *Glossina morsitans* Newst., *Bull. ent. Res.*, 52, 17-22

Nash, T.A.M., Jordan, A.M. and Boyle, J.A. (1968) The large-scale rearing of *Glossina austeni* Newst. in the laboratory. IV. The final technique, *Ann. trop. Med. Parasit.*, 62, 336-341

CULEX TRITAENIORHYNCHUS

Yoshito Wada

Department of Medical Zoology, Nagasaki

University School of Medicine, Nagasaki, Japan

Culex tritaeniorhynchus is an important vector of Japanese encephalitis, its control being of great concern to medical entomologists and epidemiologists. Various insecticides have been tested for the control of this mosquito in the laboratory and in the field, particularly in Japan and Korea. Thus, it was demonstrated by several investigators that it is difficult to control this species by the residual spraying of insecticides as a result of its exophilic characteristics (Fukui et al., 1957; Nishigaki, 1970).

For example, Nishigaki (1970) studied the change in the number of *C. tritaeniorhynchus* females after residual spraying with malathion or fenthion on the inside walls of animal sheds and dwellings, and found that whereas the females collected at animal sheds decreased in number just after spraying, they increased soon after (Figure 1). In contrast to the results with *Culex pipiens pallens*, residual spraying is ineffective in suppressing the population of *C. tritaeniorhynchus*. This is probably due to the exophilic habit of this mosquito.

On the other hand, the application of insecticides to rice fields, which are the main breeding sites of *C. tritaeniorhynchus*, is effective in controlling the mosquito population, but usually only for one or two weeks (Asahina et al., 1963; Ogata and Nakayama 1963; Suzuki and Ogata, 1961). This means that if we wish to keep the mosquito density low enough to interrupt the transmission of Japanese encephalitis throughout the epidemic season, several applications will be necessary. The effectiveness of insecticides can perhaps be prolonged when slow-release larvicides are applied, as reported by Uemoto (1969) and Wilkinson et al. (1971). Even when such larvicides are used, it will be costly to control *C. tritae-*

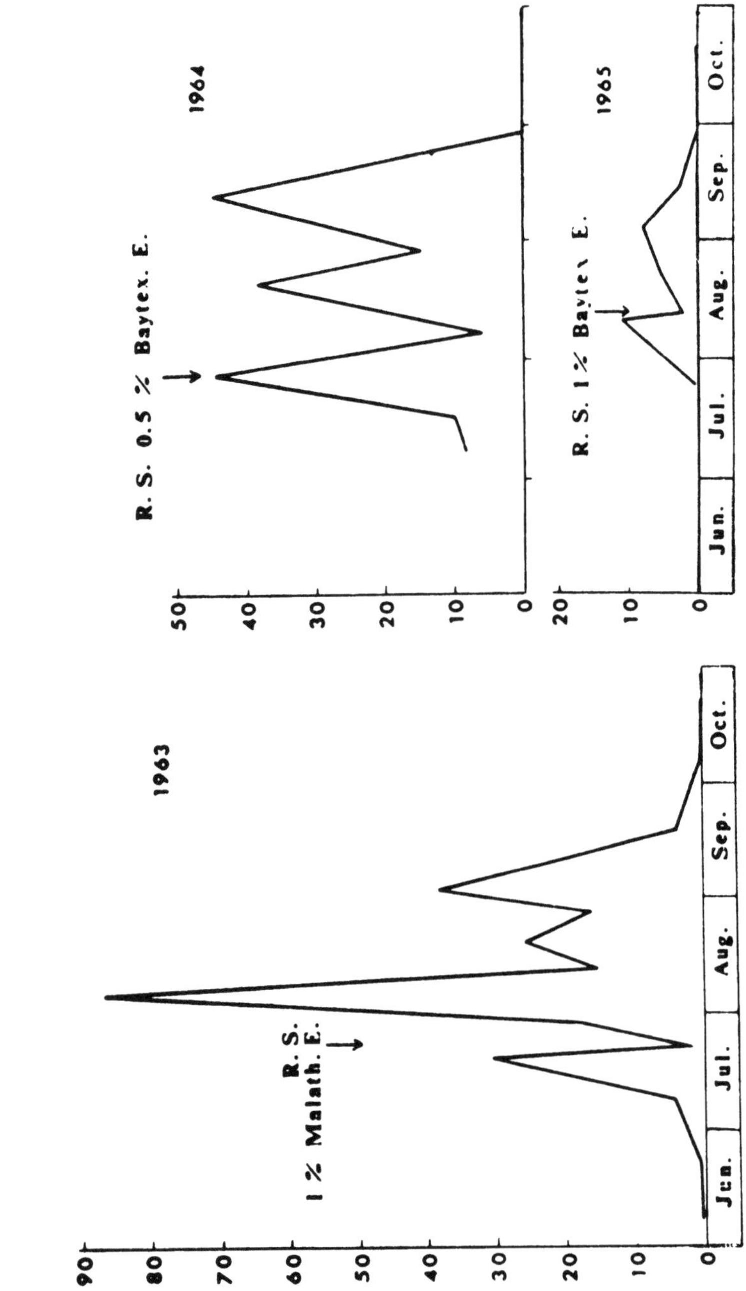

Figure 1. Effect of residual spraying (R.S.) to dwellings and cowsheds on the abundance of *C. tritaeniorhynchus* females, Matsushima, Nagasaki, (Nishigaki, 1970)

niorhynchus, because this culicine breeds over a vast area of rice fields. In agricultural spraying, insecticides are widely used to control rice pests; the population of *C. tritaeniorhynchus* is influenced at the same time. Thus, the control of this mosquito depends to a large extent on the insecticides sprayed against rice pests.

The intensive use of insecticides against rice pests has often produced a resurgence of pest populations, resistance to insecticides and residues of insecticides in the environment. For these reasons, pesticide applications in rice fields must now be modified (Sasaba and Kiritani, 1972). This change will naturally exert an influence on the populations of *C. tritaeniorhynchus*.

If we consider the fact that sufficient control of *C. tritaeniorhynchus* has never been attained to prevent the transmission of Japanese encephalitis, it is apparent that control should take a new direction. As Moore suggested in 1967, the future control of pests living in the field should be achieved by "integrated control".

Integrated control of *C. tritaeniorhynchus* may be defined as the management of the vector population with all suitable techniques to suppress the transmission of Japanese encephalitis. Sasaba and Kiritani (1972) stated rightly that when one analyses the successful cases of integrated control, it becomes evident that all were attained only after thorough ecological studies. The case of this mosquito would not be an exception.

Ecological studies of *C. tritaeniorhynchus* have been made by many investigators in various countries. Among the results reported, it is interesting to note the close relationship between changes in the mosquito population and rice culture. For example, Hill and others (1970) (Figure 2) investigated the seasonal prevalence of *C. tritaeniorhynchus* in Sarawak. A remarkable increase in the population was observed that is clearly associated with the preparation for rice planting which includes the cutting of the vegetation in fallow rice fields. The water becomes polluted with infusions from rotten vegetation, thus producing very favourable breeding sites for *C. tritaeniorhynchus*.

A similar situation is seen in other areas with a temperate climate, such as Nagasaki, Japan (Figure 3). With the exception of the over-wintering population, an increase in numbers of *C. tritaeniorhynchus* is seen at the time of transplanting, when the area of irrigated rice fields increases very greatly and when the water becomes suitable for mosquito breeding as a result of the fertilizer added for the growth of rice plants. In view of these facts, it is desirable, whenever possible, to alter the agricultural practices, including the management of water, so as to reduce the number

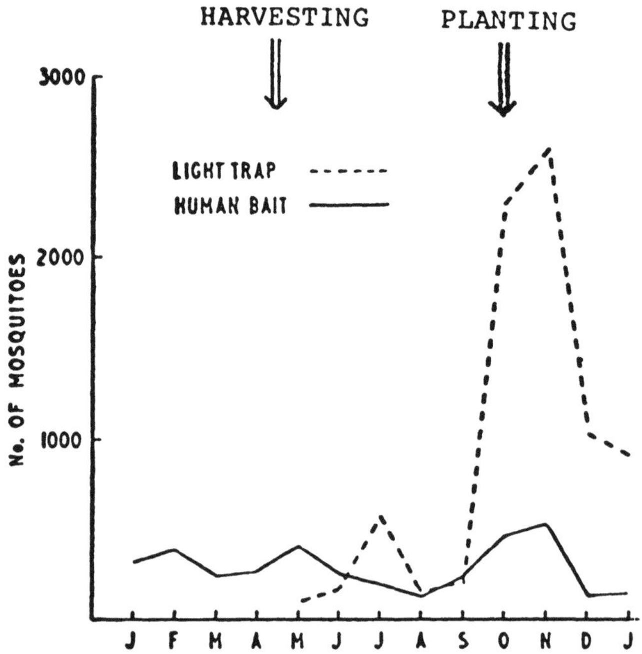

Figure 2. Seasonal prevalence of *C. tritaeniorhynchus* in relation to rice cultivation in Sarawak (from Hill, 1970).

of suitable breeding sites. However, this must be considered in relation to the production of rice; therefore this method has limitations. In the event that some changes are possible, other methods should also be considered to reduce further the number of *C. tritaeniorhynchus*.

In order to explore these methods, knowledge of the population dynamics of *C. tritaeniorhynchus* is considered very important, and such studies were begun in 1970 in Nagasaki, Japan. Although it is too early at present to draw conclusions regarding integrated control of *C. tritaeniorhynchus*, an outline of progress to date is presented below.

The study was conducted in a small farming village, Mogi, surrounded by low hills and relatively isolated from other areas (Figure 4). The area of rice fields is about 15 ha.

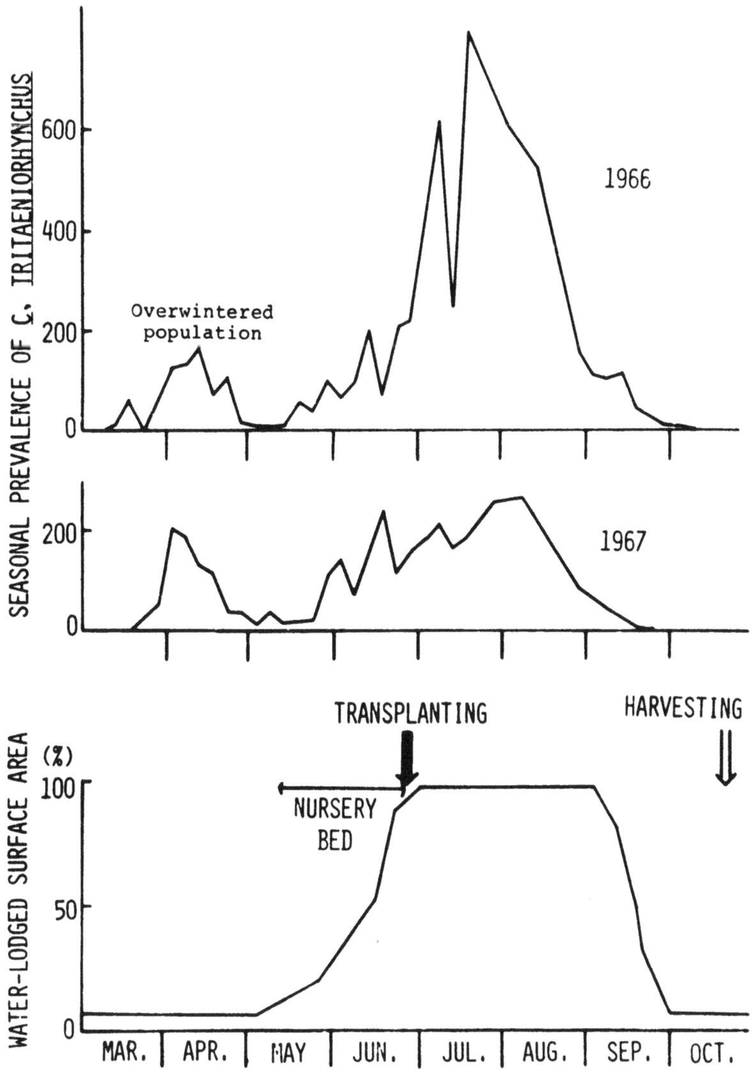

Figure 3. Seasonal prevalence of *C. tritaeniorhynchus* and water-lodged surface area of rice-field in Nagasaki, Japan

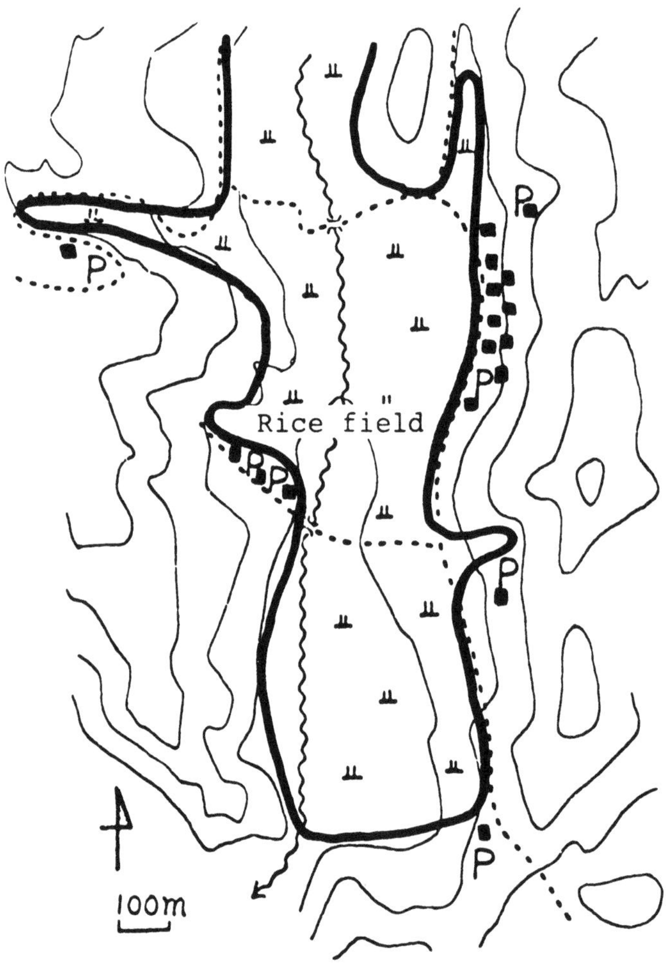

Figure 4. Map of Mogi, Nagasaki. The area of rice fields (15 ha) is surrounded by a solid line. P: piggery

In 100 rice fields, which constituted approximately one fifth of all the rice fields in the village, mosquito larvae were dipped at one-week intervals (Figure 5). The mean number of *C. tritaeniorhynchus* larvae per dip was converted to the density per square metre, using conversion factors for each larval instar and for pupae which had been determined in advance by experiments. From the density thus obtained and the area of rice fields with water, the total number of *C. tritaeniorhynchus* in each larval instar and in pupae in the whole village was obtained. If the total number of pupae is divided by the pupal period at the prevailing temperature, one can obtain an estimate for the total number of adults emerging per day in the village. The seasonal prevalence of the number of emerging adults obtained by the method mentioned above is very similar to the seasonal prevalence of females collected by a dry-ice trap. This seems to indicate that the survey of the immature stages of *C. tritaeniorhynchus* was correctly carried out.

Analyses of the data obtained for the immature stages indicate that the survival rate of larvae is generally low when the larval density is high, suggesting that some density-dependent mortality factors are involved. (Figure 6). The nature of these factors is yet to be determined.

It is apparent that some natural enemies play an important role in reducing the number of *C. tritaeniorhynchus* larvae, at least under certain circumstances. From 3,274 to 3,382 first instar larvae were placed immediately after hatching in an experimental plot of three square metres, and the number of adults emerging from them was observed in relation to natural enemies. The number of emerging adults ranged from 0 to 144, and the number of Odonata nymphs and Notonectidae nymphs and adults was larger in the plots from which fewer adults emerged. (Figure 7). Therefore, it appears certain that either the Odonata or the Notonectidae, or both, are important in reducing *C. tritaeniorhynchus* larvae. Other evidence indicates that spiders probably influence the adult population considerably. Although the density-dependency of such natural enemies is to be investigated, if even density independent, insecticides should be used in view of their influence on natural enemies. Such fish as *Gambusia* can be important natural enemies under special circumstances, but their artificial introduction into an uninhabited area does not seem to be of much practical value, at least in Japan.

The idea of suppressing a summer population of *C. tritaeniorhynchus* by the control of larvae originating from over-wintering females, of which the density is usually low, does not seem to be of operational value. Nishigaki (1970) (Figure 8) applied a larvicide every week from the early spring to all possible breeding

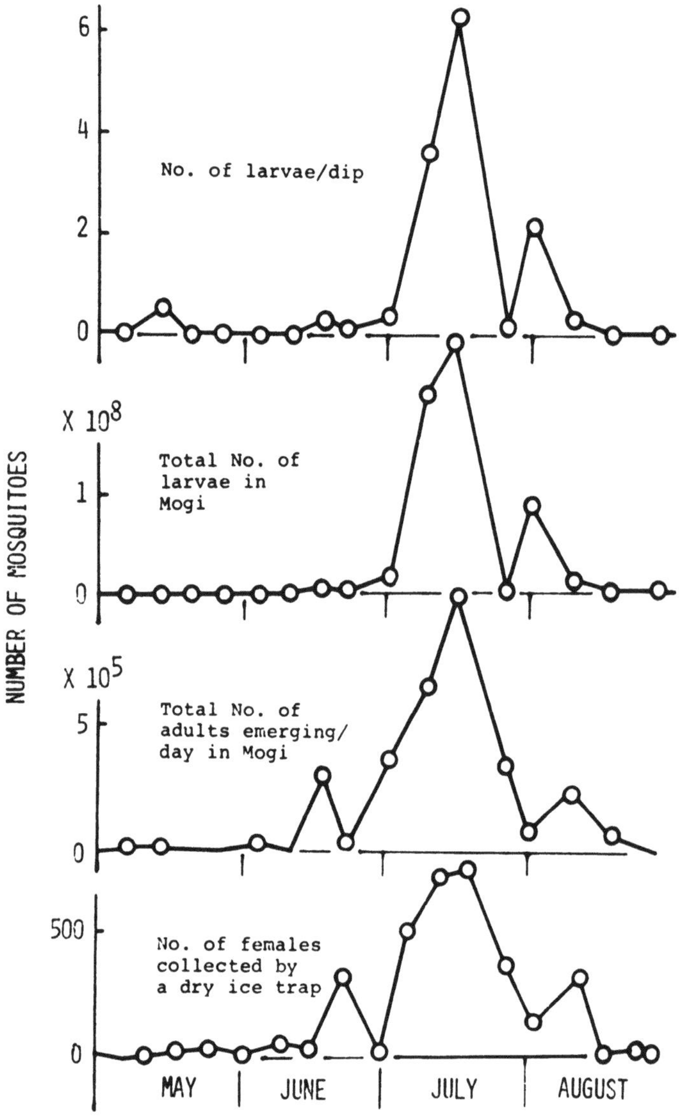

Figure 5. Seasonal prevalence of larvae and adults of *C. tritaeniorhynchus* in Mogi, Nagasaki, 1971

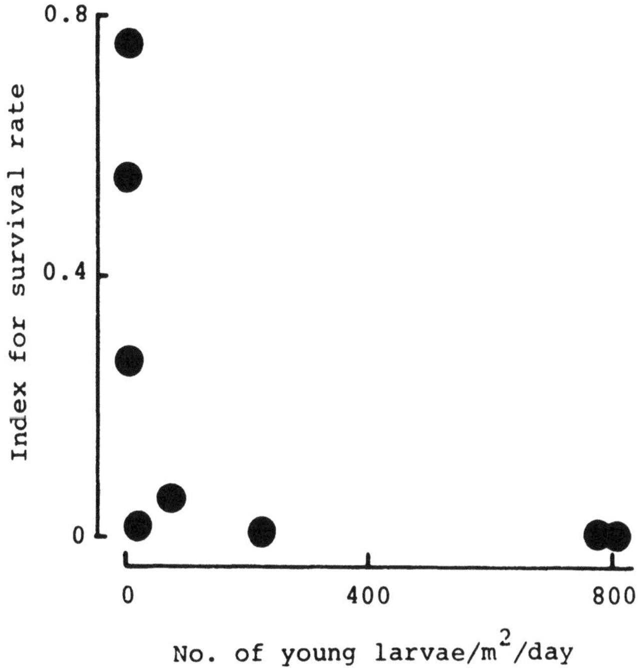

Figure 6. Relation between population density and survival of *C. tritaeniorhynchus* larvae, Mogi, Nagasaki, June-August, 1971

sites in a village. He found that the larval density could thus be kept very low until the cessation of the larvicide application, but the recovery of the population thereafter was very rapid. Taking into account that every possible effort was exerted to detect all the breeding sites and to spray the larvicide effectively, it is apparent that the pre-season control of the population is rather discouraging. Perhaps insecticides should be used only during the period when the density is to be kept low. Because of the vast number of adult mosquitos emerging from rice fields and their great potential to increase in a short time, genetic manipulation or sterilization procedures alone do not seem very promising from a practical viewpoint, at least in Japan.

However, I think that studies along these lines should be made, because they may be effectively included in the integrated control of *C. tritaeniorhynchus*, and in connection with this, I would like to emphasize again the necessity for basic ecological studies. We must consider the time most suitable for control; it is necessary to know when the epidemic starts, and therefore clarify the problem of the over-wintering of the causative virus. As for the level to

Rice field No.	No. of eggs placed	No. of adults emerged*	No. of main natural enemies		
			Odonata nymphs	Dytiscidae adults	Notonectidae nymphs + adults
1	3,274	144	108	14	0
2	3,382	14	234	12	34
3	3,279	0	583	14	40

*Estimated from the number and duration of pupae.

Figure 7. Number of *C. tritaeniorhynchus* adults emerged from the eggs placed in 3 m^2 experimental plot with different numbers of natural enemies, Mogi, Nagasaki, 1972.

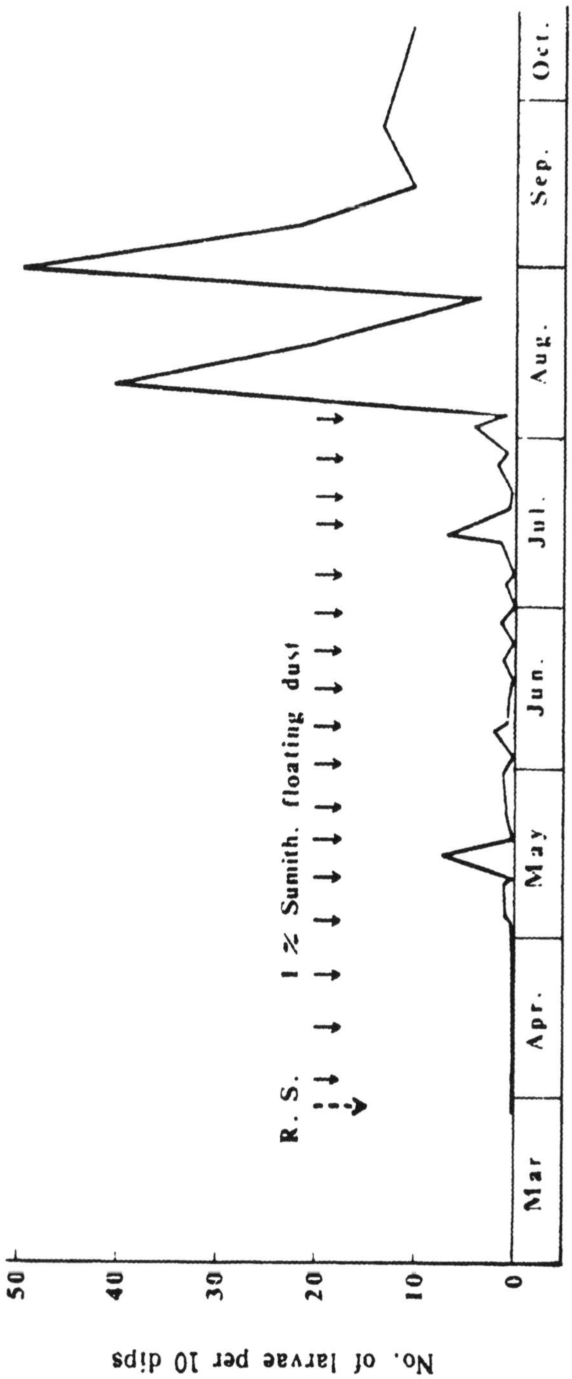

Figure 8. Effect of larvicide application to rice fields on the abundance of *C. tritaeniorhynchus* larvae, Matsushima, Nagasaki (Nishigaki, 1970)

which the population of the vector mosquito should be suppressed, a theoretical model for a Japanese encephalitis epidemic recently developed by Wada (1972) may be of value. The model indicates also the necessity of the control of the mosquito even when pigs, which are important animals amplifying the virus, are artificially immunized, because the effectiveness of pig immunization depends largely on the vector density. The role of light traps set at pigsties in lowering the transmission efficiency of the disease may have to be reassessed, in that they kill many engorged females which may include specimens recently infected by feeding on a viremic pig.

In conclusion, a practical method for controlling *C. tritaeniorhynchus* effectively enough to interrupt the transmission of Japanese encephalitis throughout the epidemic season has not yet been established, and fundamental studies on the ecology of *C. tritaeniorhynchus* are urgently required to determine the method. For the control of the mosquito, all suitable techniques must be integrated in order to reduce the mosquito population sufficiently to suppress a human epidemic of the disease. In a system of integrated control, the management of water in rice fields whenever possible and the application of insecticides at the appropriate time with little influence on natural enemies would probably be involved. It is yet to be determined whether genetic manipulation or sterilization procedures are of practical value. As for the level to which the density of the vector mosquitos must be suppressed in relation to the epidemiology of Japanese encephalitis, the theoretical model for the epidemic may be of value.

SUMMARY

Culex tritaeniorhynchus is an important vector of Japanese encephalitis, its control being of great concern to medical entomologists and epidemiologists. Various insecticides have been tested for the control of this mosquito in the laboratory and in the field, particularly in Japan and Korea. Thus it was demonstrated that it is difficult to control this species by residual spraying, as a result of its exophilic characteristics. On the other hand, the application of insecticides to rice fields, which are the main breeding sites, is effective in suppressing the mosquito population, but usually only for one to two weeks. If floating granules are used, however, the effect seems to last a little longer. This means that if mosquito density were to be kept low enough to interrupt the transmission of Japanese encephalitis throughout the epidemic season, at least several applications would be necessary. This would not only be costly, but would precipitate resistance to insecticides and an accumulation of insecticide residues in the environment. For this reason, sole reliance on pesticide applications in the rice fields must now be discontinued.

It has been suggested that the future trend in the control of pests should proceed toward "integrated control". In the case of *C. tritaeniorhynchus*, integrated control may be defined as the management of the vector population with all suitable techniques in order to prevent the transmission of Japanese encephalitis. It has also been stressed that success in integrated control can be attained only after thorough ecological studies. *C. tritaeniorhynchus* cannot be an exception. Thus, the studies on the population dynamics including the construction of life tables are now in progress in the Nagasaki area of Japan. The results obtained to date suggest that for integrated control, the management of water in rice fields and the application of insecticides should be considered in relation to their influence on natural enemies of the mosquito. Predators such as fish can be successfully used in certain circumstances. The use of genetic control or male sterilization methods do not seem very promising, because of the vast number of adult mosquitos emerging from rice fields and other breeding sites. The idea of suppressing the summer population by the control of larvae originating from over-wintering females when the density is usually not high has been considered, but it is not of much operational value, as the recovery of the mosquito population is very rapid. From the viewpoint of Japanese encephalitis epidemiology, one must consider the proper time for the control of *C. tritaeniorhynchus*. For this, it would be necessary to know when the epidemics start, and thereby try to solve the problem of the over-wintering of the causative virus. As for the level to which the population of the vector mosquito should be suppressed, a theoretical model for a Japanese encephalitis epidemic may be of value. The role of light traps set at pigsties in lowering the transmission potential of the disease may have to be reassessed, since they kill many engorged females which may include those just infected by feeding on a viremic pig, thus reducing the number of infective mosquitos.

REFERENCES

Asahina, S. et al. (1963) Control tests of *C. tritaeniorhynchus* larvae in paddy fields, *Jap. J. Sanit. Zool.*, 14, 241-244

Fukui, M. et al. (1957) On the effect of residues and DDT and dieldrin to mosquitos (studies on residual effects of insecticides 1), *Jap. J. Sanit. Zool.*, 8, 34-39

Hill, M.N. (1970) Japanese encephalitis in Sarawaki studies on adult mosquito populations, *Trans. Roy. Soc. Trop. Med. Hyg.*, 64(4), 489-496

Ogata, K. and Nakayama, T. (1963) Field tests on the control of mosquito larvae by insecticides in a paddy field, *Jap. J. Sanit. Zool.*, 14, 245-250

Moore, N.W. (1967) A synopsis of the pesticide problem, *Adv. Ecol. Res.*, 4, 75-129

Nishigaki, J. (1970) Studies on the control of *Culex tritaeniorhynchus* by larvicide application, *Trop. Med.* (Nagasaki), 11, 183-201

Sasaba, T. and Kiritani, K. (1972) Pest control in paddy fields: past, present and future, *Technocraft*, 5, 50-55

Suzuki, T. and Ogata, K. (1961) On the effect of aerial application of dipterex dust to flies and mosquitos, *Jap. J. Sanit. Zool.*, 12, 101-104

Uemoto, K. (1969) Control of *Culex tritaeniorhynchus*, *Jap. J. Trop. Med.*, 10, 111-117

Wada, Y. (1972) Theoretical model for a Japanese encephalitis epidemic, *Trop. Med.* (Nagasaki), 14, 41-54

Wilkinson, R.N., Barnes, W.W., Gillogly, A.R., and Minnemeyer, C.D. (1971) Field evaluation of slow-release mosquito larvicides, *J. econ. Ent.*, 64(1), 1-3

CULEX PIPIENS FATIGANS

M.I.D. Sharma

Director, National Institute of

Communicable Diseases, Delhi-6, India

INTRODUCTION

Culex pipiens fatigans Wiedemann, a sub-specific variant of *Culex pipiens* Linnaeus, is worldwide in distribution and is prevalent in the tropics, sub-tropics and warmer temperate climates (Horsfall 1955). *C.p. fatigans* is known to breed in waters with varying degrees of pollution, primarily soakage pits, household drains, cesspools and puddles. However, even in the absence of such water collections, the species readily breeds in clear waters such as wells and pools in river beds (Raghavan 1961).

In India, the species is widespread and has become a pest in urban and semi-urban areas. As a result of haphazard and rapid urbanization, sometimes associated with industrialization, unsanitary conditions develop owing to either an inadequate or an absence of proper drainage systems. Similar situations are being created in rural areas where a piped water supply is planned without providing for proper drainage of waste water. The case is similar in other developing countries in the tropics (Mattingly 1962; Abdulcader 1967; White 1971).

IMPORTANCE AS A DISEASE VECTOR

The importance of *C.p. fatigans* as a primary vector of bancroftian filariasis was described by Manson as early as 1878 and is widely acknowledged. The species is involved in the transmission of the classical periodic form of *Wuchereria bancrofti* in many areas of the world (Hamon and Mouchet, 1967; White 1971).

Another aspect of medical importance regarding *C.p. fatigans*, which received scant attention until recently, is its capacity to transmit arboviruses. Woodall (1965) recorded this species harbouring chikungunya virus during an outbreak in 1953 at Newala in southern Tanzania. The same virus has been isolated from pools of *C.p. fatigans* in Thailand (Hammon et al., 1960; Halstead 1963). In India strains of West Nile Virus have been isolated from *C.p. fatigans* near Poona (Maharashtra) and in North Arcot District (Tamil Nadu) (Pavri and Singh, 1965; Dandawate et al.,1969).

MEASURES FOR THE CONTROL OF *C.P. FATIGANS*

The control measures currently in use against *C.p. fatigans* are: (1) environmental sanitation; (2) the use of larvicidal oil and synthetic chemicals against its aquatic forms, and (3) the use of insecticides of vegetable or synthetic origin against imagines.

These measures individually and jointly in various combinations have been used in India and other parts of the world. The scope and limitations of each of the measures for the control of the species are briefly discussed below.

Environmental Sanitation

Control measures involving environmental sanitation include the following:

(1) filling up or elimination of ditches and depressions which are temporary or seasonal breeding places of *C.p. fatigans*;

(2) canalization of water into a lido* which can be treated with a larvicide;

(3) cleaning, desilting and deweeding of the open drains to remove stagnant water;

(4) construction of septic tanks with mosquito-proof ventilation pipes;

*A modified method of drainage, employed in "an area (that) cannot be bunded or banked to exclude flood or storm water, and then can be so dressed that all rain water will flow into a single depression, which is deepened and widened as required. All other pits are filled and the one remaining 'lido' can be periodically larvicided without much trouble or expense". (Covell 1941)

(5) underground drainage with properly maintained mosquito-proof manhole covers, and

(6) disposal of sewage and sullage water so that it does not provide breeding places.

Wherever these measures were properly undertaken, they have resulted in a reduction of the population of *C.p. fatigans*. Inadequate measures, however, such as inefficient covering of underground drains or the use of waste water for wet cultivation close to a city or town, would increase breeding sites and at the same time create problems in larviciding. Sharma (1961) outlined similar measures to the local health authorities in India for the control of filariasis. Although such measures are costly and often beyond the economic means of the developing countries, wherever possible in the long-term developmental plans for cities and towns, adequate provision should be made for their execution and maintenance.

Larval Control Measures

The application of mineral oil and insecticides constitutes the larvicidal measures generally employed. Significant impact of these measures on the population of *C.p. fatigans* is dependent on the following:

(1) the properties of the oil and sufficient persistence of the oil film on the entire water surface treated;

(2) frequency and dose of application, and

(3) extended coverage to include all breeding sites.

Usually these requisites are not easily fulfilled in large-scale operations.

In India, under the National Filaria Control Programme (NFCP) where larvicidal oil is being used extensively, the success achieved varies considerably in different places. An analysis of entomological parameters, i.e., vector density, infection and infectivity rates collected over ten years for 67 NFCP units* showed inconsistent results. Thirty-three per cent (33%) of the units demonstrated a downward trend in the indices, in 30 per cent the trend was upward, whereas the remaining units were erratic (Indian Council of Medical Research, 1971). The experience of the NFCP thus highlighted the limitations and inadequacy of larviciding operations in suppressing the *C.p. fatigans* population.

*Each NFCP unit covers an urban population varying from 20,000 to over 200,000.

The use of oil as a larvicide provides satisfactory control provided all breeding sites are covered. Among the synthetic residual insecticides used as larvicides are DDT, gamma HCH, dieldrin, malathion, diazinon, fenthion, Abate, Dursban and ethyl bromophos. In the initial stages, applications of DDT and gamma HCH were very effective in reducing the density of *C.p. fatigans*; subsequently, however, resistance to these insecticides developed (Giglioli 1948; Rajagopalan et al., 1954, 1955; Reid 1956).

Organophosphorus insecticides have not been used as yet for the control of *C.p. fatigans* larvae in India. Field trials with fenthion (Baytex), malathion, trithion and phosdrin demonstrated that these were quite effective. (Ramakrishnan et al., 1960a). Fenthion was also found to have residual toxicity for 3 days, 10 days and 15 days when applied at the rate of 0.1, 0.3 and 0.6 lb. per acre, respectively (112.5 gm, 337.5 gm and 675 gm per hectare). In Rangoon (Burma) even after five years of use as a larvicide, fenthion did not induce resistance in *C.p. fatigans*. Moreover, this insecticide was found to be cheaper than larvicidal oil (Tun and Self, 1968; Mathis et al., 1969).

Field trials in India with Abate emulsion demonstrated that its larvicidal action persisted for 10 to 36 days, 8 to 72 days and 31 to 215 days when applied at the rate of 35 ml., 70 ml., and 140 ml. of actual insecticide per hectare, respectively (Bhatnagar et al., 1969a). Dursban was also found to be an excellent larvicide. Under field conditions, residual toxicity persisted for 11 to 16 days when applied as emulsion at the rate of 28 gm, 56 gm and 112 gm per hectare of water surface. This compound, however, failed to control pupae effectively (Wattal et al., 1970a). Laboratory and field trials with ethyl bromophos when applied at the rate of 0.1, 0.2 and 0.4 ppm showed that its residual toxicity was of shorter duration indicating that it has to be applied at weekly intervals (Bhatnagar et al., 1969b). Insecticides can be used effectively as larvicides in places where oil cannot be applied, e.g., water collections in flower beds, agricultural or grass fields where insecticides with low mammalian toxicity may be used with advantage. There is, however, a great need to evaluate the potentiality and speed of selection of resistant strains to these compounds before using them on an extensive scale for the control of *C.p. fatigans*.

Use of Insecticides Against Adults

For the control of *C.p. fatigans* adults, insecticides of vegetable or synthetic origin may be used as space sprays, residual fumigants and residual insecticides.

The success of space sprays for the control of *C.p. fatigans* is dependent upon their frequency and timing. As pyrethrum has no residual effect, frequent applications are required. Therefore, it is extremely expensive as a routine measure. Residual fumigants are more convenient to use than space sprays. Limited trials with dichlorvos in Savannah, Georgia (USA) controlled *C.p. fatigans* for 14 weeks (Brooks et al., 1965). Similarly, field trials carried out with dichlorvos in India indicated effective control of *C.p. fatigans* for six to eight weeks in a hot, dry climate (Krishnamurthy et al., 1967). Residual fumigants are expensive and also less effective in open houses frequently encountered in semi-urban and rural areas. However, space sprays and residual fumigants, are useful as household insecticidal formulations for combatting the mosquito nuisance.

The application of effective residual insecticides to achieve control of *C.p. fatigans* is in rural areas and small towns, whereas in large urban areas or cities, it is not feasible because complete coverage of the indoor resting sites is not possible. Furthermore, there is always the possibility of the development of resistance to the insecticide used. It may therefore be concluded that none of the above measures can, by itself, control *C.p. fatigans* effectively enough for species eradication or filariasis control.

Insecticide Resistance in *C.p. fatigans*

Organochlorine insecticides such as DDT, gamma HCH and dieldrin have been used extensively in the field as adulticides and larvicides for the control of *C.p. fatigans*. However, the development of resistance to these insecticides seriously affected their success. The age-old method of applying larvicidal oil to breeding sites along with minor environmental sanitation measures has to be used.

In India, *C.p. fatigans* was the first species of mosquitos to develop resistance to DDT in the field (Pal et al., 1952), followed later by dieldrin and HCH (Sharma 1958; Ramakrishnan et al., 1960b). More than a decade ago, malathion and parathion were used to control this species in the United States and Japan but resistance to them also appeared (Mulla et al., 1960; Hamon and Mouchet, 1961; Barr 1962; Pennington 1968).

Studies on the spectrum of insecticide susceptibility of DDT-resistant strains revealed cross-resistance to TDE, dieldrin, gamma HCH and malathion (Pal and Kalra, 1958) to dieldrin and gamma HCH (Mohan 1960; Koshi et al., 1963), to dieldrin but not to malathion, diazinon, parathion or fenthion (Tadano and Brown, 1966). The dieldrin-resistant strain showed cross-resistance to gamma HCH, not to diazinon, fenthion and Abate. Larval selection of *C.p. fatigans* with malathion or parathion, and fenthion or diazinon, induced a slight cross-tolerance from any one organophosphorus selecting agent to all organophosphorus compounds tested and to the carbamate insecticide AC-5727. Selection with the carbamate AC-5727 demonstrated increased tolerance to the selecting agent with cross-tolerance to bromophos, gamma HCH and, in general, a loss of DDT resistance (Tadano and Brown 1966). These findings suggest that no single insecticide can remain an effective control of this species for a long period. The knowledge about the spectrum of resistance in a given area is, however, of great value in alternation or substitution of a particular insecticide for its control.

The main purpose of control of *C.p. fatigans* is to control endemic bancroftian filariasis in countries in which the problem exists both in urban and rural areas. In rural areas, adulticides are the only practical method of suppression of the vector population but none of the organochlorine insecticides can be used for long periods due to the rapid development of insecticide resistance. The organophosphate fenthion has been found to be an effective adulticide against *C.p. fatigans* for a period of about eight weeks in India (Ramakrishnan et al., 1960a), but it has to be considered that the transmission of filariasis in most areas occurs for the greater part of the year. Therefore, using adulticides of short residual toxicity would require frequent indoor spraying which is neither practicable nor economical. Furthermore, there is the problem of insecticide resistance and the toxic hazards involved in the frequent use of organophosphorus compounds.

In urban areas, because of the resistance problem, none of the organochlorine insecticides can be used, either as an adulticide or as a larvicide. Although it is difficult to provide sufficient coverage with any indoor residual insecticide in urban areas, much can be achieved by anti-larval measures. Organophosphorus compounds such as fenthion, Abate and Dursban have been found to be very effective as larvicides against *C.p. fatigans*. In Rangoon (Burma), fenthion has been used as a larvicide since 1966; as yet, no insecticide resistance in the species has been reported (Self and Tun, 1970).

PROSPECTIVE METHODS OF CONTROL

The large-scale use of insecticides has not only created resistance problems in the target species but also disturbed the balance that existed between other species and the parasite-predator complex. Increased populations of bedbugs as a result of the elimination of their natural predators resulting from DDT spraying under the National Malaria Eradication Programme in India (Wattal and Kalra, 1960) and increased breeding of houseflies in privies treated with insecticides in Savannah, Georgia (USA) (Kilpatrick and Schoof, 1956) are classic examples. However, this method of control brought out the importance of the ecosystem and genetic variability that may exist within a population. Therefore, there is a growing realisation that control measures should be applied in such a way that they cause a minimum disturbance to the ecosystem. Integrated control methods can be fitted into the prevailing ecosystem of the vector population. Such an approach would (1) avoid the loss of effectiveness of the established complexes of natural enemies and (2) combine biological genetical and environmental sanitation measures with insecticidal application. Control measures other than insecticidal and environmental sanitation are discussed below.

Biological Control

Under certain conditions such as in cesspools and pits, the use of larvivorous fish has shown encouraging results. Sasa et al. (1965) in Thailand and Dissanaike (1968) in Ceylon reported satisfactory control of *C.p. fatigans* larvae by the use of larvivorous fish known as guppies (*Lebistes reticulatus*). In India, this fish has been found breeding in a polluted waste water drain in Nagpur and maintaining effective control of the breeding of *C.p. fatigans* (Kalra et al., 1967).

Genetic Control

The basic concept of the genetic control method is to produce reproductive failure in the wild population by altering or replacing the hereditary material and releasing adequate numbers of vigorous sterile or incompatible males or by releasing self-perpetuating translocated strains. Theoretically, *C.p. fatigans* is an ideal species of mosquito for genetic manipulation as it is fairly easy to colonize and rear in large numbers. Significant advances have been made in the development of a methodology for the genetic control of *C.p. fatigans* by inducing sterility through physical (irradiation) and chemical (chemosterilization) means and by developing cytoplasmically incompatible and translocated strains.

Induction of sterility. Ionising irradiation of *C.p. fatigans* was found to induce dominant lethals which impart sterility to the insects subjected to this treatment (Ramakrishnan et al., 1962). The first field test was made in a village near Delhi (India) with males of *C.p. fatigans* sterilized by gamma irradiation. The sterile males when released induced a low degree of sterility (6%) into the field population even though the males could not compete effectively with wild males because of radiation damage (Krishnamurthy et al., 1962). However, Patterson and Sharma (1972) induced up to 25 per cent sterility into the indigenous population by the release of radiosterilized male pupae.

There are several reports of inducing sterility in the adult and pupal stages of mosquitos especially by using tepa, thiotepa, etc. (Weidhaas et al., 1961; Das 1967; Smittle et al., 1968; Pillai and Grover, 1969; Patterson et al., 1970a; Wattal et al., 1970b). The chemosterilized males are found to be as competitive as the normal males. Sharma et al. (1972) observed that no recovery of potency occurs in the chemosterilised males for a period of up to 14 days after treatment. An evaluation of thiotepa which has recently been completed has demonstrated that as a chemosterilant for *C.p. fatigans*, it has all the desired requisites.

So far, three small-scale field trials have been carried out to control *C.p. fatigans* by the release of the chemosterilized males. Varying degrees of success in these trials have been reported by Patterson et al. (1970b), Bransby-Williams (1971) and Sharma (1972) in Florida (USA), Tanzania and Delhi (India), respectively.

<u>Cytoplasmically incompatible strains</u>. The release of cytoplasmically incompatible males in a local population in a circumscribed area is known to reduce the insect population (Pal 1966; Laven 1967a). The phenomenon of a blockage of the sperm before karyogamy due to cytoplasmic factors has been found in allopatric populations of *C. pipiens* over the range of its distribution (Laven 1957). Laven (1967b) has recorded 17 different crossing types known in this species complex. Laven (1967a) released incompatible males in a village near Rangoon (Burma) and claimed successful eradication of the *C.p. fatigans* population in a 12-week period. Krishnamurthy and Laven (1972) have prepared incompatible and reserve strains for use in the field in the Delhi (India) area.

<u>Translocated strains</u>. Certain chromosomal aberrations such as translocations and pericentric inversions are known to produce inheritable semisterility. Laven (1969) and Laven and Jost (1971) in *C. pipiens pipiens* and McDonald and Rai (1970) in *A. aegypti* produced radiation-induced translocated strains. The mosquitos heterozygous for reciprocal chromosomal translocation if mated with normal mosquitos would produce a 50 per cent reduction of fertility. The double translocation hetrozygotes are more useful for population suppression, producing approximately an 80 per cent reduction of fertility.

Laven and Aslamkhan (1970) have put forward the concept of the replacement of the indigenous population of *C.p. fatigans* with an integrated strain. The integrated strain has both bi-directional cytoplasmic incompatibility with the local strain and high lethality from a translocation. The release of such a strain is possible in the egg-raft stage, which is both economical and convenient. The strain, theoretically, after eradicating the local population, exterminates itself because of low fertility.

INTEGRATED CONTROL SYSTEM

The different control methods discussed above have their own limitations. It appears that none of them alone will be capable of controlling the vector population. There is an urgent need to utilize control measures in the field that are harmonious with the prevailing ecosystem. The use of larvicides along with measures aimed at improving environmental sanitation appears to be the best combination for the effective control of *C.p. fatigans* in urban areas. For rural areas, indoor residual spraying of houses with synthetic insecticides is the only method available at present. Biological control techniques are yet in a rudimentary stage, and whereas genetic control may provide valuable additional methods for controlling the species, at present they too have limitations for wide-scale use in the field.

From the foregoing, it appears that whereas the approaches for integrated control are apparent, our knowledge of the ecology of the target species and its ecosystem is too meagre to exploit fully the possibilities that exist in initiating an integrated control system. There is a need for greater recognition of the problems involved amongst health workers to meet this challenge.

SUMMARY

An assessment of past and future measures for the control of *C.p. fatigans* demonstrates that the existing measures *viz.*, environmental sanitation, larvicides and adulticides,are largely inadequate.

In view of the appearance of strains of *C.p. fatigans* resistant to organochlorine and organophosphorus insecticides,the use of genetic and biological methods of control has been discussed. Whereas these measures show promise, they also have limitations for wide-scale use. Therefore, an integration of various methods and their judicious use appears to be the most suitable technique.

REFERENCES

Abdulcader, M.H.M. (1967) The significance of *Culex pipiens fatigans* Wiedemann problem in Ceylon, *Bull. Wld Hlth Org.*, 37, 245-250

Barr, A.R. (1962) Proceedings and papers of the 13th annual conference of the California Mosquito Control Association Inc. Visalia, Calif, 88-103

Bhatnagar, V.N., Joshi, G.C. and Wattal, B.L. (1969a) Laboratory and field evaluation of Abate (o,o,o',o' - tetramethyl o,-o' -

Thiodi-Para - Phenylene Phosphorothioate) in the control of *Culex pipiens fatigans* Wied. larvae, *J. Com. Dis.*, 1, 203-215

Bhatnagar, V.N., Joshi, G.C., Dhar, S.K., Basu, P.C. and Wattal, B.L. (1969b) Trials with organophosphate larvicide. Ethyl bromophos (0.0. diethyl, 0-2-5-dichloro-4 bromophenyl thionophosphate against *Culex fatigans* Wied. larvae, *Ind. J. Ento.*, 31, 13-20

Bransby-Williams, W.R. (1971) A field release of male *Culex pipiens fatigans* sterilized by apholate, *East Afr. Med. Jour.*, 48, 68

Brooks, G.D., Carmichael, G.T., Schoof, H.F. (1965) Evaluation of large-scale treatment with dichlorvos for the control of *Culex pipiens quinquefasciatus* Say *Mosquito News*, 25 427-431

Covell, G. (1941) *Malaria control by antimosquito measures*, Thacker, Spink & Co., Calcutta

Dandawate, C.N., Rajagopalan, P.K. Pavari, K.M. and Work, T.H. (1969) Virus isolations from mosquitos collected in North Arcot District, Madras State, and Chittor District, Andhra Pradesh between November, 1955 and October, 1957, *Ind. J. Med. Res.*, 57, 1420-1426

Das, M. (1967) Sterilization of *Culex pipiens fatigans* Wiedemann by Apholate *Bull. Wld. Hlth Org.*, 36, 949-954

Giglioli, G. (1948) Malaria, Filariasis and Yellow Fever in British Guiana, Medical Deptt. - Mosquito control service

Halstead,S.B., Yamarat, C. and Scanlon, J.E. (1963) Thai haemorrhagic fever in Thailand epidemic of 1962, *Jour. Med. Asso.* (Thailand), 46, 449-462

Hamon, J. and Mouchet, J. (1961) La résistance aux insecticides chez les insects d'importance médicale. Methodes d'étude et situation en Afrique au sud du Sahara (The resistance to insecticides of insects of medical importance in South Africa of Sahara) *Med. Trop.*, 21, 565-596

Hamon, J. and Mouchet, J. (1967) La résistance aux insecticides chez *Culex pipiens fatigans* Wiedemann, *Bull. Wld Hlth Org.*, 37, 277-286

Hammon, W.M., Rudnick, A. and Sather, G.E. (1960) Viruses associated with epidemic haemorrhagic fevers of the Philippines and Thailand, *Science*, 131, 1102-1103

Horsfall, W.R. (1955) *Mosquitoes, their bionomics and relation to diseases*, The Ronald Press Company, New York

Indian Council of Medical Research (1971) Assessment of the National Filaria Control Programme (India) 1961-1970, *Tech. Rep. Ser. No. 10.*

Kalra, N.L., Wattal, B.L. and Raghavan, N.G.S. (1967) Occurrence of larvivorous fish *Lebistes reticulatus* (Peters) breeding in sullage water at Nagpur-India, *Bull. Nat. Soc. Mal. Com. Dis.*, 4, 253-254

Koshi, T. Dixit, R.S. and Perti, S.L. (1963) The selection of insecticide resistant strains of the mosquito, *Culex fatigans* Wield. in the Laboratory, *Ind. J. Mal.*, 17, 23-31

Kilpatrick, J.W. and Schoof, H.F. (1956) Fly production in treated and untreated privies, *Publ. Hlth. Rep. (Wash)*, 71, 787-796

Krishnamurthy, B.S., Kalra,R.L. and Sharma, S.K. (1967) Field evaluation of 'Dichlorvos' for the control of mosquitoes, *Bull. Ind. Soc. Mal. Com. Dis.*, 4, 23-33

Krishnamurthy, B.S., Ray, S.N. and Joshi, G.C. (1962) A note on preliminary field studies on the use of irradiated males for reduction of *C. fatigans* Wied. populations, *Ind. J. Mal.*, 16, 365-373

Krishnamurthy, B.S. and Laven, H. (1972) Preparation of a cytoplasmically incompatible strain of *Culex pipiens fatigans* for use in genetic control in India, WHO/VBC/72.389

Laven, H. (1957) Vererbung durch Kerngene and das Problem der ausserkaryotischen Vererbung bei *Culex fatigans* 1, Kernvererbung, *Z. Verebungel*, 88, 443-477

Laven, H. (1967a) Eradication of *Culex pipiens fatigans* through cytoplasmic incompatibility, *Nature*, 216, 383-384

Laven, H. (1967b) Speciation and evolution in *Culex pipiens*. In *Genetics of insect vectors of disease*, Wright, J.W. and Pal, R. (eds) Elsevier, Amsterdam, 251-307

Laven, H. (1969) Eradicating mosquitoes using translocations, *Nature*, 221, 958-959

Laven, H. and Aslamkhan, M. (1970) Control of *Culex pipiens pipiens* and *C.p. fatigans* with integrated genetical systems, *Pakistan Jour. Science*, 22, 303-312

Laven, H. and Jost, E. (1971) Inherited semisterility for control of harmful insects. I. Production of semisterility due to translocation in the mosquitoes, *Culex pipiens* L., by X-rays, *Experientia*, 27, 471

Manson Patrik (1878) On the development of *Filaria senquinus hominis* and on the mosquito considered as nurse, *J. Linn. Soc. Zool.* (London), 14, 304-11

Mathis, H.L., Graham, J.E., Abdulcader, M.H.M. and Self, L.S. (1969) Relative cost of oil and fenthion for larvicidal control of *Culex pipiens fatigans* Wiedemann in Rangoon, Burma, WHO/VBC/69-138 (Unpublished document)

Mattingly, P.F. (1962) Population increase in *Culex pipiens fatigans* Wiedemann, *Bull. Wld. Hlth Org.*, 276, 579-584

McDonald, P.T. and Rai, K.S. (1970) *Aedes aegypti:* Origin of a new chromosome from a double translocation heterozygote, *Science*, 168, 1229-1230

Mohan, B.N. (1960) A note on DDT-resistant *Culex fatigans* with cross-resistance to gamma BHC and dieldrin, *Ind. J. Mal.*, 14, 175-178

Mulla, M.S., Isaak, L.W. and Axelrod, H. (1960) Laboratory and field evaluation on new insecticides against mosquito larvae, *Mosquito News*, 20, 256-261

Pal, R. (1966) Genetic control of vectors of diseases with special reference to *Culex pipiens fatigans*, WHO/VBC/66.198

Pal, R. and Kalra, R.L. (1958) Comparative susceptibility of DDT-resistant and normal *Culex fatigans* larvae to diphenyl methane derivatives, chlorinated hydrocarbons and organophosphorus insecticides, *Ind. J. Mal.*, 12, 439-445

Pal, R., Sharma, M.I.D. and Krishnamurthy, B.S. (1952) Studies on the development of resistant strains of houseflies and mosquitoes, *Ind. J. Mal.*, 6, 303-316

Patterson, R.S. and Sharma, V.P. (1972) Induction of sterility in a field population of mosquitoes *(Culex fatigans)* using irradiated males, WHO/VBC/72.339 (Unpublished document)

Patterson, R.S., Ford, H.R., Lofgren, C.S. and Weidhaas, D.S. (1970a) Sterile males: their effect on an isolated population of mosquitoes, *Mosquito News*, 30, 23-27

Patterson, R.S., Weidhaas, D.E., Ford,H.R. and Lofgren, C.S. (1970b) Suppression and elimination of an island population of *Culex pipiens quinquefasciatus* with sterile males,*Science*, 168, 1368-69

Pavri, K.M. and Singh, K.R.P. (1965) Isolation of West Nile Virus from *Culex fatigans* mosquitoes from Western India, *Ind. Jour. Med. Res.*, 53, 501-505

Pennington, N.E. (1968) Resistance of *Culex tritaeniorhynchus* Giles and *Culex quinquefasciatus* Say to malathion on Okinawa with notes on susceptibility to other insecticides, *Mosquito News*, 28, 193-198

Pillai, M.K.K. and Grover, K.K. (1969) Chemosterilization of *Culex pipiens fatigans* Wiedemann by exposure to aquatic stages. I. Sterilization potential of certain aziridinyl compounds, *Bull. Wld. Hlth. Org.*, 41, 915-928

Raghavan, N.G.S. (1961) The vectors of human infections by Wuchereria species in endemic areas and their biology, *Bull. Wld. Hlth. Org.*, 24, 177-195

Ramakrishnan, S.P., Sharma, M.I.D. and Kalra, R.L. (1960a) Laboratory and field studies on the effectiveness of organophosphorus insecticides in the control of *Culex fatigans*, *Ind. J. Mal.*, 14, 545-566

Ramakrishnan, S.P., Raghavan, N.G.S., Krishnaswami, A.K., Nair, C.P., Basu, P.C., Singh, D., and Krishnan, K.S. (1960b) National Filaria Control Programme in India - A review (1955-59), *Ind. J. Mal.*, 14, 457-508

Rajagopalan, N., Vedamanickam, J.C. and Ramni, S.R. (1954) Development of resistance to BHC by *C. fatigans* in the course of a larval control programme, *Bull. Nat. Soc. Ind. Mal. Mos. Dis.*, 2, 211-213

Rajagopalan, N., Vedamanickam, J.C. and Ramani, S.R. (1955) Tolerance of larvae of *Culex fatigans* Wiedemann to Gammexane applied as a larvicide at Kambakonam, South India, *Ind. J. Ent.*, 12, 159-164

Reid, J.A. (1956) Field trials of larvicides for use against *Culex pipiens fatigans* with a note on the life-cycle of this mosquito, *Ann. Trop. Med. Parasit.*, 50, 129-136

Sharma, M.I.D. (1958) Laboratory and field development of resistance in mosquitoes, *Ind. J. Mal.*, 12, 401-411

Sharma, M.I.D. (1961) The Kerala State Municipal and Corporation Health Inspectors and Health Assistant Conference, 2nd State-Conference 1961, Trivandrum

Sharma, V.P. and Patterson, R.S., Seetharam, P.L. and LaBrecque, G.S. (1972) Radiosterilization of the tropical house mosquito *Culex fatigans* Wied.: laboratory and field cage studies, WHO/VBC/72.345 (Unpublished document)

Sharma, V.P. (1972) Personal communications

Sasa, M. Kurihara T., Dhamvanij, O. and Harinasuta G. (1965) Studies on mosquitoes and their natural enemies in Bangkok. III. Observations on a mosquito-eating fish "guppy" *Lebistes reticulatus* breeding in polluted waters, *Japan J. Exp. Med.*, 35, 63-80

Self, L.S. and Tun, M.M. (1970) Summary of the field trials (1964-1969) in Rangoon, Burma using organophosphorus larvicides and oils against *Culex pipiens fatigans* in polluted water, WHO/VBC/70.187 (Unpublished document)

Smittle, B.J., Mount, G.A., Das,M. and Rajapaksa, N. (1968) Apholate and gamma irradiation compared as sterilants for *Culex pipiens quinquefasciatus* Say (Diptera, Culicidae), *Mosquito News*, 28. 201-204

Tadano, T. and Brown, A.W.A. (1966) Development of resistance to various insecticides of *Culex pipiens fatigans* Wiedemann. *Bull. Wld. Hlth. Org.*, 35, 189-201

Tun, M.M. and Self, L.S. (1968) The susceptibility of *Culex pipiens fatigans* larvae to fenthion after two years of larval control in Rangoon, Burma, WHO/VBC/68.77 (Unpublished document)

Wattal, B.L., Bhatnagar, V.N., Joshi, G.C. and Sharma, S.K. (1970a) Laboratory and field evaluation of Dursban (O,O-diethyl-O-3, 5-6 trichloro-2-pyridyl phosphorothioate) as a larvicide against *Culex pipiens fatigans* Wied. larvae, *J. Comm. Dis.*, 2, 53-60

Wattal, B.L., Das, M. and Raghavan, N.G.S. (1970b) Place of chemosterilized male technique in the control system of *Culex fatigans*, *J. Comm. Dis.*, 2, 80-86

Wattal, B.L. and Kalra, N.L. (1960) *Pyralis pictalis* Curt. (Pyralidae: Lepidoptera) larvae as predators of eggs of bed bugs *Cimex hemipterus* Fab. (Cimicidae; Hemiptera), *Ind. J. Mal.*, 14, 77-79

Weidhaas, D.E., Ford, H.R., Gahan, J.B. and Smith, C.N. (1961) Preliminary observations on chemosterilization of mosquitos, *Proceedings 48th Ann. Meeting, N.J. Mosq. Exterm. Association*, 106-109

White, G.B. (1971) The present importance of domestic mosquito (*Culex pipiens fatigans* Wiedemann) in East Africa and recent steps towards their control, *East, Afri. Med. J.*, 48, 266-274

Woodhall, J.P. (1965) Summary of virus isolations from arthropods at this institute to the end of 1964. In East African Virus Research Institute Report, July 1963 to December 1964

Additional reference:

Dissanaike, A.S. (1968) Proc. Seminar 3rd S.E.S. Reg. Meet. Parasit. Trop. Med., p. 149

Ramakrishnan, S.P., Krishnamurthy, B.S. and Ray, S.N. (1962) Laboratory studies on the use of irradiated sterile males to reduce *C. fatigans* Wied. populations, *Ind. J. Mal.*, 16, 357-364

SUBJECT INDEX

MIX
Papier aus verantwortungsvollen Quellen
Paper from responsible sources
FSC® C105338

If you have any concerns about our products,
you can contact us on
ProductSafety@springernature.com

In case Publisher is established outside the EU,
the EU authorized representative is:
Springer Nature Customer Service Center GmbH
Europaplatz 3, 69115 Heidelberg, Germany

Printed by Libri Plureos GmbH
in Hamburg, Germany